Register No

THE NATIONAL SCHEDULE OF RATES

INFORMATION FOR PURCHASERS

The National Schedules are part of a complete option for undertaking term maintenance contracts. Purchasers rest in the knowledge that assistance is available from consultants who are all able, from experience in the field of measured term contracting, to assist users in the setting up and operation of this type of contract.

The National Schedules are fully resourced, updated and incorporate new items annually. They are providing an increasing number of owners and contractors with an efficient, consistent and accountable system for the commissioning of building repairs and maintenance works.

The cost of undertaking repairs and maintenance work can be estimated in advance of orders being placed. A further benefit to building owners and occupiers lies in the fact that standards of materials and workmanship can be firmly controlled and is likely to be of higher quality than the work valued on a daywork basis.

Apart from the printed version, the major National Schedules are available in a data format with integrated software. Samples of which can be downloaded from our website www.nsrm.co.uk

As a purchaser, you will receive the benefit of being able to share the experiences of other purchasers in operating measured term contracts by registering in the space provided below.

Those wishing to receive further detailed advice and assistance should contact the Administrator on: 01296 339966, or by fax 01296 338514 or email: nsr@nsrmanagement.co.uk.

It will help us to help you if you complete and return the information below:

REGISTRATION OF PURCHASERS - 08/09   No: ☐   [from back of book]

Name: _____

Organisation: _____   Job title: _____

Address: _____
_____
_____

Telephone No: _____   Fax No: _____

email address: _____

How long have you been using NSR? _____

Named contractor using NSR on your behalf _____

NSR Management
Pembroke Court
22-28 Cambridge Street
Aylesbury
HP20 1RS

Affix
stamp
here

The Administrator
NSR Management
Pembroke Court
22-28 Cambridge Street
Aylesbury
HP20 1RS

NATIONAL SCHEDULE of Rates 2008-2009

Published and Compiled Annually by:

NSR Management Ltd

General, Publications, Data Licence and Software available direct from:

NSR Management Ltd
Pembroke Court
22-28 Cambridge Street
Aylesbury HP20 1RS

Tel: 01296 339966
Fax: 01296 338514
www.nsrm.co.uk

Mechanical Services in Buildings

Part 1

Introduction
Guidance to Users
Preliminaries/General Conditions
Materials and Workmanship Preambles
Measurement and Pricing Preambles

Part 2

Compiler's Notes
Schedules of Descriptions and Prices
Schedule of Basic Prices

ISBN 978-1-904739-37-1

Whilst all reasonable care has been taken in the compilation of the Mechanical Services in Buildings Schedule the Co-authors and NSR Management will not be under any legal liability of any kind in respect of any mis-statement error or omission contained herein or for the reliance any person company or authority may place thereon.

Important

© All rights reserved. No part of any of the National Schedule of Rates may be reproduced stored in a retrieval system or transmitted in any form or by means electronic mechanical photocopying recording or otherwise without the prior consent of the Copyright owners.

This book is sold subject to the condition that it shall not, by way of trade or otherwise, be lent, resold, hired out or otherwise circulated without the publisher's prior written consent in any form of cover or binding other than that in which it is published and without a similar condition including this condition, being imposed upon the subsequent purchaser.

MECHANICAL

PART 1

Allow us to help you with the Property Maintenance Jigsaw

- Contract Advice
- QS Services
- Benchmark Services
- Condition Surveys
- DDA Surveys
- Management of Contracts

One STOP Shop

QS Services, Benchmarking, Condition Surveys

For over 12 years, NSR Management have been promoting and developing the National Schedule of Rates on behalf of the Co-authors, the Society of Construction and Quantity Surveyors and the Construction Confederation. During this time the portfolio of National Schedule of Rates have become synonymous with the management of Measured Term Contracts, and from 2006 NSR Management will compile the National Schedules.

Together with the appropriate contract conditions, these National Schedules allow customers to issue a series of works orders, confident that the charges for the work will be based on a pre-determined and agreed basis of measurement and pricing. Over the years the National Schedules have saved time and money for hundreds of organisations just like yours.

We like to think we listen carefully to our customers needs, and in response to their requests NSR Management now offer a comprehensive **One STOP Shop** consultancy service for those organisations involved in term contracting. Our team have over 70 years experience in the construction industry and we thought it time to let our customers have the benefit of this expertise.

Just look at some of the ways in which our comprehensive consultancy service can help you:

- Comprehensive advice on all aspects of Term Contracting
- Efficient preparation of Tender Documents to suit clients needs
- Sound Technical Advice
- Thorough Benchmarking and value for money assessments
- A wide range of other QS Services
- Smooth management of Measured Term Contracts
- Effective Stock Condition and DDA Surveys

NSR Management can offer flexible cost effective solutions to all of your property management needs. Combine this with the range of National Schedule of Rates, including our new web based schedules, and it is not hard to see why we lead the way...

Please contact us for more information

NSR Management

NSR Management Limited
Pembroke Court
22-28 Cambridge Street
Aylesbury, Bucks HP20 1RS

tel: 01296 339966
fax: 01296 338514

nsr@nsrmanagement.co.uk

www.nsrm.co.uk

'Our business is your property'

General Index

Page

Part 1

Section A	: Introduction	1/1
Section B	: Contract Applications	1/5
Section C	: General Guidance to Users	1/7
Section D	: The Advisory Panel	1/13
Section E	: Tender Forms and Contract Documents	1/15
Section F	: Preliminaries/ General Conditions	1/19
Section G	: Material and Workmanship Preambles	1/39
Section H	: Measurement and Pricing Preambles	1/63

Part 2

Index	2/1
Summary of Amendments	2/5
Compiler's Notes	2/6
Schedule of Description and Prices	2/17
Basic Prices: Labour, Plant and Materials	BP

W T PARKER LIMITED

ELECTRICAL & MECHANICAL TERM CONTRACTORS TO DEFENCE ESTATES AND THE NHS

W.T. Parker Ltd
24-28 Moor Street
Burton on Trent
Staffordshire
DE14 3SX
Tel: 01283 542661
Fax: 01283 536189
www.wtparker.co.uk

SECTION A

INTRODUCTION

The National Schedule was first introduced by the Co-Authors - the Society of Chief Quantity Surveyors and the Construction Confederation in 1982 for use on maintenance and minor building work to meet the competition requirements of the Local Government Planning and Land Act 1980.

This schedule of rates for Mechanical Services in Buildings has a similar format to that used for the National Schedule of Rates for Building Works, the prices shown against the description of work having been analysed and set down into their respective elements of labour, plant and materials. However the measurements of works, is defined in the item description, the measurement preambles and the Compiler's Notes.

The users attention is drawn to the materials prices used in this document. Wherever possible, the costs of materials are based on standard published price lists current at April/May 2008. No discounts have been allowed against the materials costs and users are advised to check with their local suppliers to ascertain material costs and make due allowance in their tender percentage adjustment for such materials against those contained in the National Schedule.

The labour rates in this schedule include adjustments to the annual awards announced by the Construction Industry Joint Council and by the Joint Industrial Boards (JIB) for Plumbing & Mechanical Engineering Services and for the Electrical Contracting Industry. Detailed calculations of all allowances added to the basic rates of wages for craftsmen and labourers are given in the Compiler's Notes in Part 2 of the Schedule.

Whereas there is a requirement that all parties to the contract are required to comply with all relevant legislation current at the time of execution of the Works, users attention is particularly drawn to the 'Landfill tax', these regulations came into force on 1 October 1996, and were revised with effect from 1 April 1999, increasing the cost for disposal of active waste, the costs of which are to be included in the tendered percentage adjustment to schedule rates. (see page 1/9 for details of other percentage adjustment costs.)

The rates contained in this schedule do not include for any costs associated with 'The Construction (Design and Management) Regulations 1994'. Although it is envisaged that the majority of works undertaken will not require compliance with the CDM regulations, we would recommend that Clients appoint a CDM Co-ordinator and ensure that the tendered percentage adjustment includes for complying with the CDM Regulations and all as detailed in the approved Code of Practice – managing construction for Health and Safety, Construction (Design and Management) Regulations 1994.

This will ensure that a framework for complying with the CDM Regulations is in place and that the Contractor will be recompensed for his costs.

The rates contained in this Schedule do not include for any costs associated with working with asbestos. When carrying out work on asbestos based materials the contractor shall comply with the Control of Asbestos Regulations 2002 and appoint a licensed Specialist Subcontractor. Where minor work with asbestos is encountered and which can be carried out by a contractor who does not normally require an HSE licence as defined in the Approved Code of Practice and Guidance published by the Health & Safety Executive, the contractor may carry out the work as requested by the Contract Administrator at agreed rates or on a daywork basis [see page 1/26]

As an increasing number of property managers in both public and private sectors are forced to review their costs of building maintenance and repair they are considering the benefits to be gained from using a measured term contract, a comprehensive schedule of rates and a computerised management system such as is available through the National Schedule.

The value of the National Schedule to any public or private property owner or estate manager is best understood by considering how it may be used and the advantages it has over alternative methods of commissioning and paying for maintenance and minor works.
The use of the pre-priced and widely accepted National Schedule together with appropriate Contract Conditions allows the Client to issue a series of works orders, confident in the knowledge that the charges

for the work when carried out will be made on a predetermined and agreed basis of measurement and pricing. Without the National Schedule the client either has to produce his own schedule, or invite a series of competitive quotations for the works before placing orders, or issue orders and subsequently be charged on a lump sum or daywork basis without prior knowledge of the end cost. In the latter case there may be no clearly defined contract conditions covering such matters as insurance, payment, or standards of workmanship and materials.

The published National Schedule data is produced and stored using computer software and is available under a licence to subscribers for use on their computer systems. For further details contact the Administrator on 01296 339966.

The Form of Tender is available from NSR Management Ltd. The JCT Measured Term Contract, together with Practice Note and Guide is available from NSR Management, RIBA, RICS etc. Standard Method of Measurement exception clauses are incorporated in the Preambles where appropriate.

The use of the National Schedule and the JCT Form of Measured Term Contract has been explained at conferences across the country organised by the Society of Construction and Quantity Surveyors (formerly Society of Chief Quantity Surveyors), Local Government Associations and other organisations.

Other seminars covering the use of the National Schedule and the importance of Energy Efficient Design of buildings have taken place on a number of occasions in the South of England, promoted by Southern Electricity.

HISTORY

The need for a National Schedule for local authority maintenance and other building work arose from the change in requirements governing Local Government direct labour accounting brought about by the Local Government Planning and Land Act 1980.

In its Accounting Code of Practice for Direct Labour Organisations, the Chartered Institute of Public Finance and Accountancy recommend the use of schedules and this has been substantiated by the demands made by a large number of authorities. Statistics published by CIPFA show an increased use of schedules of rates in local authorities since 1981.

David Hannent, a Chief Quantity Surveyor, examined those schedules which were available at the beginning of 1982 and found that none fitted the needs of his authority. The Society of Chief Quantity Surveyors thereupon agreed to his proposal that they should adopt the development of a National Schedule. Discussions were held with a number of other local authority organisations and with the Construction Confederation.

The Construction Confederation had meantime produced a schedule and both the Construction Confederation and SCQS felt that their respective schedules had features worth adopting. It was agreed to merge the two documents and Contract Conditions, Form of Tender and Standard Minute Values were introduced at a later date, enabling users to invite tenders and analyse costs by means of a comprehensive range of documents.

A major change was introduced in 1990 to revise the descriptions, format and coding to comply with SMM7 and to introduce the JCT 1989 Conditions of Measured Term Contract with an updated Model Form of Tender. The Society of Chief Quantity Surveyors also published a revised Code of Procedure for Building Maintenance Works using the National Schedule in November 1997.

In 1991 the National Schedule published schedules of rates for Electrical and Mechanical services in Buildings and entered into a joint venture with the PSA on their updated schedule of rates for Landscape Management. A further joint venture led to the publication in 1994 of the National Roadworks Schedule of Rates. In 1995 the National Housing Maintenance Schedule was first published. This is a composite schedule of rates of approximately 400 items using the Building Works Schedule as a basis of pricing. In 2001, following customer feedback the Schedules for Mechanical Services in Buildings and Electrical Services in Buildings were comprehensively reviewed, updated and upgraded.

CHARACTERISTICS

The characteristics of the resultant National Schedule are:-

(i)   Each rate is broken down into its component labour, materials and plant and may form the basis for a bonus measurement or material allocation.

(ii) Being produced and published by the Co-Authors as a national document, no extras apart from the annual cost of each edition will fall upon users.

(iii) It is published as a schedule of rates including other relevant information e.g. Tender Forms, Contract Conditions, Preliminaries and Preambles, lists of basic prices of labour, plant and materials.

(iv) It complies with the Standard Method of Measurement 7th Edition (SMM7 amendment 3), except for minor variations specifically referred to in the Preambles section, so that measurement disputes are minimised.

(v)   It may be used for works by private contractors as well as for DLO use.

(vi) It is computer based and, since it has a library of headings and descriptions by which it is produced, considerable further development is possible. The library has been structured so that new items may be inserted into their appropriate places allowing original items to retain their existing numbers.

(vii) It is comprehensive and although originally designed for public sector maintenance and repair work, it nevertheless may easily be adapted for other purposes. The schedules have been successfully used for fast-track rehabilitation contracts to reduce pre-contract periods.

(viii) It is acceptable to the industry as a whole and its flexibility allows continuous development.

(ix) An Advisory Panel comprises representatives of interested professional organisations and, being a fully constituted, independent body, provides the Co-Authors with advice and comment on any revisions of the National Schedule thought to be desirable.

(x)   A Standard Minute Value Schedule is also available in a separate volume which gives an analysis of the labour content of each item.

(xi) Single trade schedules may also be produced from the available data. 'The Painting and Decorating Schedule of Rates' is currently available in this format.

(xii) It is sponsored by the Construction Confederation and the SCQS as the Co-Authors.

The Society of Construction and Quantity Surveyors and the Construction Confederation recognise the need for updating and improving any document of this kind and welcome suggestions and comments from subscribers. Technical queries and suggestions addressed to the Technical Secretary are referred where appropriate to the Advisory Panel and the Compilers and have led to the introduction of new items and other improvements in this and previous editions.

SECTION B

CONTRACT APPLICATIONS

The National Schedule has been priced from basic principles and is reviewed annually.

Each item in the Schedule is calculated using the appropriate materials prices, plant hire rates and all-in labour hourly rates as shown in the List of Basic Prices which is included in Part 2.

Tables showing the basis of the calculations of the all-in labour rates and a list of firms from whom material prices are obtained are included in the Compiler's Notes in Part 2.

The rates shown in the List of Basic Prices are nationally based and, by the use of percentage adjustments to the material, labour and plant values, the Schedule may be applied to any part of the United Kingdom. The rates do not however include profit and overheads and other costs which are noted in the contract conditions and preliminaries and should be allowed for as part of the tendered percentage adjustment to the rates.

As the National Schedule is divided into sections it may be used for tendering:-

(i) As a complete schedule

(ii) In groups of sections

(iii) In individual sections

It may also be used with the following types of contract:

(i) Fixed price

(ii) Fluctuating - based upon changes in the National Schedule which is up-dated and re-published annually

(iii) Fluctuating - other method, by agreement

It may be used to tender for:

(i) All work in any geographical area

(ii) Work in any particular type of building

(iii) Work on an estate

(iv) Term contracts

(v) Lump sum contracts

With particular regard to its use by Direct Labour Organisations it must be emphasised that the National Schedule is aimed at providing a tender document based on current market prices. It may be used to provide cost targets after suitable adjustments. An Authority will be able to test the viability of its Direct Labour Organisation provided that a recent competitive tender, based on the National Schedule, has been obtained for work similar to that carried out by the DLO.

The National Schedule **with its adjusting percentage** gives the market value but the cost to users will be determined by their own accounting methods, usually by work-study-based bonus systems. In any case, it is necessary to link the National Schedule to the system and this may be possible, especially if the bonus system follows an established order. Any such set of bonus descriptions, when built up from or broken down to the National Schedule, will show an area of 'best fit'. The rogue items may then be priced separately to provide a direct local comparison.

SECTION C

GENERAL GUIDANCE TO USERS

Schedules of rates have been widely used for term contracts for maintenance and repair work for many years. The National Schedule is suitable for use with measured term contracts which include both reactive or day-to-day maintenance as well as pre-planned repair and improvement works. If the tender documents indicate that programmed cyclical maintenance is to be included it is probable that the tendered percentages will be lower as a result and it will also save the client the necessity of pre-pricing and inviting a number of individual tenders and contracts. Such works should not subsequently be taken out of the measured term contract to achieve more competitive bids for the remaining reactive maintenance and repair work.

Term Contracts

Where an employer wishes to engage a contractor to carry out ad hoc maintenance and repair work and also requires competitive quotations for the work, experience has shown that term contracts based on a priced schedule of rates may be suitable. However the following factors should be noted:-

(a) Within certain broad parameters, such as overall value, location and time, the term contract is for an unspecified amount of work at an unspecified location at an unspecified time.

(b) This type of contract is intended to cope with large numbers of small orders placed over the whole term of the contract and this allows the statistical "swings and roundabouts" to operate to reduce the overall pricing risk to reasonable proportions.

(c) In arriving at his percentage on or off the schedule of rates the contractor will have to price what he thinks will be a representative selection of items of work to be done. The tender invitation should include a clear description of the types of work to be done and the pricing restricted to those sections of the schedule covering the intended work. For example it would be wrong to invite tenders for work which is described as predominantly plumbing and heating work and issue substantial orders for replacement of gutters. Such requirements for unrepresentative work could seriously affect the pricing calculations and could not be held to come within the normal "swings and roundabouts" of the system.

(d) Both contractors and employers need experience with a schedule system before prices can be stabilised. Users are therefore reminded that they should look upon the introduction of a schedule of rates as a long term project and ensure that they have adequate technical support services in their organisation.

(e) Contract conditions are specifically designed for this type of work.

Single Projects of Refurbishment or Modernisation Work

Where a single substantial package of work can be identified it might be advantageous to tender separately on the basis of the schedule. In these circumstances more specific information may be given about scope, timing and location of the work and the contractor will price with greater confidence. For example, the kind of project envisaged for this approach might be an estate of houses or a school or college which are to be modernised and redecorated. The description of the work would be backed up by drawings, specifications, schedules and project preliminaries.

The contract conditions for this type of project could be carried out under the JCT 98 or 2005 Forms with approximate quantities where the quantities are priced at National Schedule rates to which tenderers will apply a percentage adjustment.

Where it is not possible to measure in sufficient detail and the Measured Term Contract is used, it is nevertheless recommended that some pre-measurement or assessment is undertaken to evaluate the scope of the contract to attract the best tenders.

Single New Work Projects

As explained above the more specific the description of the work, the less the risk in pricing for both contractor and employer. It is possible to fully describe new work through drawings, specifications and bills of quantities, which may be priced relative to the National Schedule if desired.

Selection of Contractor

The Form of Tender for use with Measured Term Contracts let on the National Schedule provides for tenderers to insert a percentage adjustment to the rates listed in the National Schedule for work to be measured and valued in accordance with the contract and for other percentage adjustments to the prime cost of labour, materials, plant and subcontractors for work carried out or valued under the contract as daywork.

It is recommended that the client informs the tenderers of the value of dayworks that will be assessed in the selection process as different combinations may give different results.

In the table, contractor C is the lowest price by virtue of the more competitive dayworks percentages, after allowing for the assessed values of measured work and dayworks, as is seen from the calculation of anticipated final account costs.

TABLE

Contractor	PERCENTAGES IN TENDER			
	Percentage Adjustment	DAYWORK		
		Material	Plant	Subcontract
A	+10	20	15	15
B	+11	15	10	10
C	+12	5	5	5

ASSESSMENT

If, for example, the client estimates that in every £100,000 of the final account £85,000 will be valued at schedule rates or pro rata, and that there will be; £5,000 labour; £4,000 materials; £3,000 plant and £8,000 sub-contract values in dayworks, the estimated final account values of the three tenders will be:-

Net Value of Work			Final Account Values		
			Contr. A	Contr. B	Contr. C
		£	£	£	£
Schedule		80,000	88,000	88,800	89,600
Dayworks -	Labour	5,000	5,000	5,000	5,000
	Materials	4,000	4,800	4,600	4,200
	Plant	3,000	3,450	3,300	3,150
	Sub-Contr	8,000	9,200	8,800	8,400
Total		100,000	110,450	110,500	110,350

This Assessment assumes a contract where little overtime will be instructed by the Contract Administrator. Where a contractor is expected to carry out a significant part of the contract outside normal working hours an allowance in respect of overheads and profit on non profit overtime rates [Percentage adjustment of Clause 5.7 of the JCT Measured Term Contract] should be incorporated into the tender assessment.

Another factor recommended to clients who order a regular amount of this work is to maintain a separate select list of contractors for schedules. In this way the contractor may establish a workforce with knowledge and experience of schedules, have surveyors and managers who understand the system and the client's requirements, and thus provide the most competitive price.

Certainly, contractors are now available who specialise in National Schedule work and these should be sought. The other advantage in having a specialist list is that it gives the inevitable "swings and roundabouts" a chance to work.

It is significant that tenders are now stabilising as experience of methods and pricing levels is obtained. Encouragement and stimulation given to 'schedule keen' contractors will provide an increasing pool of knowledge which will benefit all parties and is certain to reduce costs and prices in the long term.

Tendered Percentage adjustment

The tendered percentage is deemed to include but not limited to:-

Contractor's overheads and profit
Travelling time where applicable
Permit to work
Setting out and checking for dimensions
Off-loading, manhandling and transport and delivery of materials and plant
Square cutting and waste where different from allowance included in rates
All necessary security precautions
Protection, drying and cleaning – see item in Section F on page 1/27
Compliance with Regulations
Safety Health and Welfare regulations and policies
CDM Regulations
Landfill tax
Control of noise
Any site establishment costs necessary including temporary telephones
Rubbish removal including tipping
Management and staff
Disbursements arising from the employment of work people
Service and facilities
Hand tools mechanical plant where not included in Part 2
Temporary works
Scaffolding –   temporary staging up to 1.5 metres and ladders for access to buildings of not more than 2 storeys in height
Client training
Adjustment to standard specification requirements notified to tenderers and/or variations for local conditions
Records and drawings
Identification and labelling of services
Builders work information (rates are provided for chases and holes etc)
Work in conjunction with ceiling installations
Painting and priming of services
General bonding and earthing
Inspection testing and commissioning of the works (see also references to testing on pages 1/33 & 1/40)
Complying with preliminaries particulars, conditions of contract and preambles
Trade discount adjustment

Specialist Goods and Work not specified within the Schedule

There will be some items of work which are not suitable for inclusion in a standard schedule of rates because the circumstances on which they are required and consequently the cost will vary considerably e.g. asbestos removal etc. There are other non-standard items which are required only rarely. Sometimes it will be better for a client to have separate contracts for such works, however where these works are included as non-standard items on orders given to the Term Contractor the following procedures are to be adopted.

The Contractor is entitled to a 5% cash discount on the nett cost of specialist goods and a 2.5% cash discount on the nett cost of specialist work. Appropriate adjustments to achieve these percentages will be made when compiling the final account.

The nett cost of specialist goods and work will NOT be subject to any tendered percentage (omission or addition) but the Contractor is entitled to a percentage addition for overheads, profit and general attendance on the cost of specialist goods and work **after** the adjustment to provide cash discount described above. Specialist attendance shall be measured separately in accordance with the provisions of the Standard Method of Measurement.

The nett cost is defined as the cost charged to the contractor after deduction of all trade or other discounts except for cash discounts referred to above.

NOTE: Where items containing Prime Cost sums have an identical fix only item then the materials used for the fix only item will NOT be treated as specialist goods but adjusted for waste factors, cash discounts and the tender percentage (omission or addition) as for Prime Cost items.

Prime Cost Items specified within the Schedule

All Prime Cost Items will be adjusted in the final account. The Contractor is entitled to a 5% cash discount on the nett cost of goods supplied under instruction against a Prime Cost Item specified within the Schedule. Appropriate adjustment to achieve this percentage will be made when compiling the final account.

The effect of the adjustment of the cost of the Prime Cost Item (after the adjustment to provide cash discount described above) will be subject to any tendered percentage (omission or addition).

The nett cost is defined as the cost charged to the contractor after deduction of all trade or other discounts except for cash discounts referred to above.

No waste factors have been allowed relative to prime cost items.

Overtime Payments

Overtime payments are to be paid in accordance with Clause 5.7 of the Standard Form of Measured Term Contract which refers to Working Rule 4, published by Construction Industry Joint Council, which states when overtime payments should commence.

The rates for overtime should be paid as follows:-

Mondays - Fridays

First four hours	-	Time and a half
After first four hours until normal starting time next morning	-	Double time

Saturdays

First four hours	-	Time and a half
After first four hours	-	Double time

Sundays

All hours	-	Double time

© NSR 2008 - 2009

Notwithstanding the requirements of the Standard Form an easy administrative way of dealing with out of hours work is to adopt the measures set out in the following example.

Work carried out being valued using Schedule Rates:-

Example

	£	£
S1703/195		16.95
Add tendered percentage Adjustment say 10%		1.69
		18.64
Add non productive overtime (time and a half)		
50% of labour £9.62 x 50%	4.81	
Add tender percentage adjustment say 15%	0.72 +	5.53
		£ 24.17

Work carried out being valued using Daywork Rates:-

Example

	£	£
3 hours @ £25.00		75.00
Add non productive overtime (double time)		
100% x 3 hours @ £25.00	75.00	
Add tender percentage adjustment say 50%	37.50 +	112.50
		£ 187.50

This example has been calculated in accordance with the JCT Measured Term Contract utilising the Definition of Prime Cost of Building works of a Jobbing or Maintenance Character which allows other basis to be used where appropriate to the class of labour concerned.

Procedures

Utilising the documents requires the client to think carefully about procedures and staffing, especially in the context of term contracts. This can only be done at local level but is preferably defined in the tender documentation. No national system can be devised to satisfy all requirements, and the National Schedule is not a panacea for this headache. The matters of progress of job tickets, priorities, emergency work, valuations, dayworks, supervision and access are all specific to the individual client. Clauses resolving these items may be inserted in the preliminaries or within any relevant explanatory procedure document. Suggested headings for preliminaries clauses are given elsewhere (see Section F).

Appendix A of Section F contains a list of supplementary clauses, which a client may wish to include in contract documentation dependent on particular circumstances. Advice on the incorporation of these clauses and other more particular advice on the contents of Conditions of Contracts and Preliminaries is available from NSR Management on a consultancy basis.

User Feedback

In order to continuously update and improve the Schedules, feedback is required from users and clients alike. Please return the User Response Form issued with this documentation to the Administrator, NSR Management, Pembroke court, 22-28 Cambridge Street, Aylesbury, HP20 1RS.

SECTION D

THE ADVISORY PANEL

The panel has been established as an independent body providing a forum for discussion and feedback to the Co-Authors.

The Panel's role is seen to be one which provides users with an independent committee capable of directing the needs of the user. The Co-Authors are committed to providing the Panel with the facilities to act as such.

Members of the Advisory Panel represent the following bodies concerned with the execution of construction and maintenance work:-

1. British Constructional Steelwork Association Ltd

2. British Gas

3. Construction Confederation

4. Chartered Institute of Public Finance and Accountancy

5. Chief Building Surveyors Society

6. Society of Chief Architects

7. Society of Construction and Quantity Surveyors

8. Society of Electrical and Mechanical Engineers in Local Government

SECTION E

TENDER FORMS AND CONTRACT DOCUMENTS

The Form of Tender and the Contract Conditions are available as separate publications. The Forms of Tender are available from NSR Management Ltd. and the Contract Conditions are published by RIBA Publications and available through their usual outlets and NSR Management Ltd.

It is recognised that some alterations may be made in applying these forms to particular contractual requirements, but as printed they provide a model format suitable for most requirements.

In view of the variety of possible combinations of tender the following must be considered as a guide. It is possible to tabulate the requirements of a tender thus:-

(a) **Which sections of the National Schedule will apply?**

(b) Are dayworks to be appended?

(c) How is scaffolding to be paid for?

(d) How is travelling to be paid for?

(e) How is plant to be paid for?

(f) How are the CDM regulations to be paid for?

(g) What is the definition of the area of operation of the contract?

(h) How is emergency work to be paid for?

(i) What is duration of the contract?

(j) Is the contract fixed price or fluctuating?

(k) How is the work to be measured?

(l) What type of work is to be executed?

(m) What preliminaries and preambles apply?

(n) What comprises the Contract Documents for tendering?

NOTES

(a) **Which sections will apply?** - Rather than have a tender which has a different percentage against each section it is advised that a single percentage is applied.

(b) **Dayworks** - Daywork should be calculated in accordance with JCT MTC clauses 5.4

(c) **Scaffolding** - Further details on this may be found in the Preliminaries in Section F; rates for scaffolding are included in Part 2 (Schedule of Descriptions and Prices).

The rates contained within this schedule, together with the 'Percentage A' adjustment include for working within high rise buildings and for working from scaffolding up to a height of 10 metres above ground level.

Where external operations are to be carried out from scaffolding or other methods of access above 10 metres the tender documents should give specific information on how such orders are to be executed and, where not deemed included in the 'Percentage adjustment, allow for an additional percentage adjustment to the rates depending on the height eg:

    10 - 15 metres 2.5%
    15 - 20 metres 3%
    20 - 25 metres 3.5%
    25 - 30 metres 4%

(d) **Travelling** - The cost of this item should be included in the tender percentage adjustment. Where the properties to be maintained are spread over a wide area this could be a significant overhead cost and relevant information should be given to the tenderers. Some contracts provide for adjustments to tendered percentage to allow for different distances travelled to jobs.

(e) **Plant** - The cost of plant such as compressors, kango hammers, drills etc. is included where appropriate and identified in the break down of the rates shown in Part 2. If the client intends to make some plant available to the contractor it should be clearly identified in the preliminaries giving details of the terms under which it may be used.

(f) **CDM Regulations** - The Client should appoint a CDM Co-ordinator and ensure the Principal Contractor prepares a construction phase plan. The cost to the Contractor in complying with the CDM Regulations should then be included in the tender percentage.

(g) **Definition of area** - The tenderer must be advised of the location and conditions affecting his work. For example, an 'area' could be an estate, a building or a geographical area.

(h) **Emergency Work** - During normal working hours emergency work may be paid for under the provisions of clause 5.8 of the JCT Measured Term Contract. Such work outside normal working hours may be paid at daywork rates and may involve a minimum call-out charge. The tender documents should give specific information on how such orders will be executed and paid for and what levels of priority will be used. Priorities may be listed as footnote (U) to Contract Appendix item 4.

Further requirements for the provision of emergency services may be given as follows:

Emergency Services

The contractor must provide and maintain an out-of- hours emergency service to avoid danger to the health and safety of residents and the public or serious damage to buildings and other structures. The service must be provided 24 hours each day including at weekends, and during holiday periods.

The names, addresses and telephone numbers of suitable competent persons who may be contacted outside working hours shall be provided by the contractor. The telephone must be manned and not an answering machine or answering service or mobile service.

(i) **Duration** - The exact dates for commencement and completion of the contract must be defined.

(j) **Fixed price or fluctuating** - Alternative methods may be adopted.

For fixed price work it is only necessary to identify the rates which will apply by reference to a particular edition of the schedule.

For fluctuating contracts, which are expected to be for periods over one year duration, it is recommended that a fixed date be specified when the rates will be varied, i.e. all work executed after 1st August will be priced at that schedule rate until 31st July the following year.

It may be more convenient to apply a change date to the date the work is measured but this must be agreed by the user.

(k) **How is the work to be measured?** - The work may be measured by the Contract Administrator; alternatively the contract provides for the Contractor to carry out the appropriate measurement for subsequent verification of the account by the Contract Administrator. Where the client intends that the contractor shall measure the works in whole or in part this is to be stated in item 9 of the Appendix to the contract.

(l) **What type of work?** - The National Schedule does not define what information must be given at tender stage about the type of work involved and the range expected, eg. "all repairs to heating installations". Further information may be given as follows:-

"The scope of work comprised in any order or in the whole contract cannot be pre-determined and no undertaking is given regarding continuity or overall value of the work and the contractor must allow in his tender for all intermittent or abnormal workloads.

As an indication of the anticipated scope of the work the following information is given as a guide:

No. of properties..............
No. of orders per week...............
Average value of order..............

Average classification of work:-

Space heating	....%
Gas services	....%

Work in other trades will also be required."

(m) **Preliminaries and Preambles** - Model Preliminaries and Standard Preambles are included as part of the Schedule. It is important that the Contract Administrator should define the conditions and the character of work as carefully as possible at tender stage and the circumstances in which it is executed.

(n) **The Contract Documents for Tendering** - Contractors prepared to tender will be expected to subscribe to the National Schedule of Rates. Letters of invitation from the Client will then be required which will incorporate the tender documentation set out below.

Recommended Tender Documentation from the Client:

This would include:

(i) **Letter of invitation**

(ii) **Tender Form**

This must note the Contract Conditions if different from those included in the schedule of rates.

(iii) **Schedule of Works**

The Model Preliminaries allow for Description of Works, however, as much detail as possible of the actual work to be done should be provided and this is best achieved by way of an accurate Schedule of Works. The Schedule of Works may be the job ticket(s) issued during the contract, a prepared Schedule of the Works actually required, or an overall definition of the work to be done when and if circumstances arise. As much information as possible at tender stage will assist in obtaining a relevant and competitive tender.

(iv) **Preliminaries & Preambles**

The relation of all documents to one another is most important and any adaptation of Model Preliminaries and Standard Preambles should be done with care to ensure that they are compatible. The appendix to the contract must be completed by the client.

(v) **Drawings**

Any drawings that are available will assist the tenderer.

(vi) **Other information**

All attached correspondence and information should relate to the main documents and indicate the exact content of the Tender and Contract Documentation. The date and time for receipt of Tenders and a plain addressed envelope for their return to the client should be provided.

Further information on the procurement and administration of term contracts is contained in a Code of Procedure for Building Maintenance Works published jointly by the Society of Construction and Quantity Surveyors and NSR Management.

SECTION F

PRELIMINARIES/GENERAL CONDITIONS

The following information is given to provide general and particular information relating to the proposed contract which could affect the tender and to assist in the completion of the appendix to the JCT Measured Term Contract Conditions pages 5 - 10.

Appendix A contains a list of supplementary clauses which may be applicable on a particular contract dependant on circumstances.

CONTENTS

	Page
PROJECT PARTICULARS	1/20
DRAWINGS	1/20
THE SITE/EXISTING BUILDINGS	1/20
DESCRIPTION OF THE WORK	1/20
THE CONTRACT/SUB CONTRACT	1/21
EMPLOYERS REQUIREMENTS	1/21
: Tendering/Sub letting/Supply	1/21
: Provision, content and use of documents	1/22
: Management of the Works	1/22
: Quality Standards/Control	1/22
: Security/Safety/Protection	1/24
: Specific limitations on method/ sequence/timing	1/28
: Facilities/Temporary work/ Services	1/30
: Operation/Maintenance of the finished works	1/32
CONTRACTOR'S GENERAL COST ITEMS	1/32
WORK/MATERIALS BY THE EMPLOYER	1/33
NOMINATED SUB CONTRACTORS AND SUPPLIERS	1/33
WORK BY STATUTORY AUTHORITIES	1/33
APPENDIX A – LIST OF SUPPLEMENTARY CLAUSES	1/35
APPENDIX B – GENERAL ADVICE ON TUPE	1/37

PROJECT PARTICULARS

Employer : Name and address of Employer to be given

Location : The site(s) is/are situated *

Access : Provision for access will be arranged by the Contract Administrator in accordance with clause 3.4 of the Contract Conditions, access to the site(s) is by .... *

The Contractor is deemed to have visited the site(s) and ascertained the means of access and any limitations thereto: no claims for additional costs caused by access difficulties arising from the execution of the works and which were identified at the date of tender will be allowed.

The Contractor is to agree access to the Site(s) with the Contract Administrator before commencement of the works.

DRAWINGS

The following drawings will form part of this contract:

Drawing No	Description
....	....
....	....

Further contract drawings may be issued during the term of the contract.

THE SITE/EXISTING BUILDINGS

Site boundaries: Where works are to be carried out on open sites the boundaries and other details must be provided with the order.

Existing buildings: Work to existing buildings should be clearly defined by room or dwelling number or drawing as necessary.

Existing mains or services: Contractors attention should be drawn to services in the vicinity of proposed works and any limitations regarding their use or protection should be given.

DESCRIPTION OF THE WORK

General Description: The works covered by this contract and these schedules of rates comprise ....

.... *

* Contract Administrator to complete/delete as applicable

THE CONTRACT/SUB-CONTRACT

Conditions of Contract: The Form of Contract will be the JCT Measured Term Contract 2006 Edition with amendments unless otherwise stated. NOTE – This contract makes no provision for Nominated Suppliers or Sub-Contractors.

(Where an alternative form of contract is selected a schedule of the clause headings should be given).

Special conditions or amendments to standard conditions should be stated.

All insertions to the contract appendix are to be made including employers insurance responsibility.

Performance Guarantee Bond

Note This may not be required for measured term contracts where completion of works ordered occurs regularly throughout the term of the contract. Similarly a retention percentage of stage payments is not needed for the same reason.

EMPLOYERS REQUIREMENTS

Tendering/Sub-letting/Supply:

Details to be given of any restrictions or requirements e.g.

MATERIALS to be supplied by client.

TENDERING to be in accordance with the principles of the Code of Procedure for Single Stage Selective Tendering 1996 (Alternative 1 or 2).

AMENDMENTS: No amendments to be made to the schedules or other contract documents by tenderers. Tenders containing amendments or qualifications may be rejected.

COSTS TO BE INCLUDED: The contractor is to allow for costs of fulfilling all liabilities and obligations referred to in the preliminaries, preambles and other tender documents as part of his tendered percentages.

FIXED PRICE: All works carried out under the contract shall be valued in accordance with the published edition of the schedule of rates named in the tender and contract subject to the tendered percentage adjustment *

or

FLUCTUATING PRICE: All works carried out under the contract shall be valued in accordance with the edition of the schedule of rates current at the date the work is carried out subject to the tendered percentage adjustment.*

(Note this does not apply to dayworks)

TRANSFER OF UNDERTAKINGS (PROTECTION OF EMPLOYMENT) REGULATIONS 1981 AND THE ACQUIRED RIGHTS DIRECTIVE 1977: The requirements of these regulations may or may not* apply to this contract and professional advice should be sought. General advice is contained in Appendix B.

---

* Contract Administrator to complete/delete as applicable.

Provision, content and use of documents

All tender documents may be inspected by appointment during normal office hours at the employer's offices situated at*

....

....

Tel..

Tenderers are recommended to obtain their own copy of the National Schedule which includes the Conditions of Contract, Preliminaries, Preambles together with the schedule of descriptions and prices.

SCHEDULE COMPLIES WITH SMM7: the Schedule of Rates has been prepared generally in accordance with the Standard Method of Measurement of Building Works, 7th Edition (amendment 3), subject to amplifications inherent in any of the descriptions therein. Items or methods of measurement contrary to the Standard Method of Measurement of Building Works are stated in the Preambles relating to the Schedule of Rates.

PREAMBLES QUALIFY DESCRIPTIONS: items in the Preambles relating to the Schedules of Rates are deemed to qualify and to be part of every description to which they refer.

RECIPROCAL USE OF RATES; rates may be used reciprocally between sections of the Schedules of Rates in the settlement of accounts.

Management of the Works

PERSON IN CHARGE: During the carrying out of the works, the Contractor is to keep on the works a competent person in charge who shall be empowered to receive and act upon any instructions given by the contract administrator or his representative.

Quality Standards/Control

QUALITY CONTROL: the contractor is to establish and maintain procedures to ensure that the works including the work of all subcontractors, comply with specified requirements. Maintain full records, keep copies for inspection by the Contract Administrator and submit copies on request. The records must include:-

a) The nature and dates of inspections, tests and approval
b) The nature and extent of any nonconforming work found
c) Details of any corrective action

SETTING OUT: the Contractor is to take the dimensions from existing premises and check with dimensions given on the drawings. Allow for setting out the works and providing all instruments and attendance required for checking by the Contract Administrator.

MATERIALS, LABOUR AND PLANT: provide all materials, labour and plant and all carriage, freightage, implements, tools and whatever else may be required for the proper and efficient execution and completion of the works. All materials are to be new unless otherwise specified.

OFF-LOADING AND MANHANDLING: the Contractor shall allow for all off-loading and manhandling into position including placing into and removing from temporary site storage prior to final positioning, all general materials, plant and items of equipment.

* Contract Administrator to complete/delete as applicable.

SAMPLES AND STANDARDS OF MATERIALS: the Contractor shall allow for obtaining samples of materials as required by the Contract Administrator. Such samples to be approved by the Contract Administrator before use or application in the works. All material subsequently used in the works is to be of equal quality in all respects to the approved sample.

MANUFACTURER'S RECOMMENDATIONS: means the manufacturer's recommendations or instructions, printed or in writing current at the date of tender.

COMPLIANCE WITH REGULATIONS: the Contractor shall allow in his tendered percentage Adjustment to the schedule of rates for and ensure that the works and components thereof comply with all recommendations, requirements and current editions of the following:-

a) British Standard Specifications

b) British Standard Codes of Practice

c) The Factories Act

d) The Health and Safety at Work Act

e) The Gas Act and Regulations

f) Electricity at Work Act and Regulations

g) The Regulations for Mechanical Installations issued by the Heating, Ventilation and Air Conditioning Association

h) The Regulations for Electrical Installations issued by the Institution of Electrical Engineers

i) The Electrical Supply Regulations

j) The Rules and Regulations of the Local Electricity, Gas and Water Authorities

k) Fire Precaution Act and Regulations and the requirements of the Local Fire Authorities

l) Portable Appliance Regulations

m) Regulations relating to the control of Asbestos and Legionnaires Disease

n) Regulations relating to Waste Management

PROPRIETARY NAMES: the phrase 'or other approved' is to be deemed included whenever products are specified by proprietary name.

INCOME AND CORPORATION TAXES ACT 1988 or any statutory amendment or modification thereof: the attention of the Contractor is drawn to this Act which replaces the Finance (NO.2) ACT 1975. The provisions of this Act are explained in the Board of Inland Revenue pamphlets.

The Contractor is also reminded that it is his duty and responsibility to satisfy himself that all Sub-Contractors are approved by the Contract Administrator and hold an appropriate Sub-Contractor's Certificate from the Inland Revenue.

DIMENSIONS: dimensions stated or figured dimensions on the drawings are to be adhered to. Any discrepancy between the drawings is to be brought to the notice of the Contract Administrator for clarification and instruction.

GENERALLY: all standards referred to within these documents shall be held to be the latest edition published at the date of tender. A reference to any Act of Parliament, Regulation, Code of Procedure or the like shall include a reference to any amendment or re-enactment of same.

CO-ORDINATION OF ENGINEERING SERVICES: The site organisation staff must include at least one person with appropriate knowledge and experience of mechanical and electrical engineering services to ensure compatibility between engineering services, one with another and each in relation to the works generally.

REINSTATEMENT: the Contractor shall make good and re-instate in working order any existing security system switched off, damaged or otherwise rendered inoperable by the works.

Security/Safety/Protection

SECURITY PRECAUTIONS: the Contractor is to allow for any security precautions that may become necessary in relation to the adjoining properties during the course of the works and is to allow for adequate measures to prevent access from scaffolding or similar means.

The Contractor shall issue ID badges to all employees who are engaged on the Works and these badges are to be displayed at all times. Contractor employees shall wear suitable clothing to identify them as employees of the Contractor.

SAFEGUARDING THE WORKS, MATERIALS AND PLANT AGAINST DAMAGE AND THEFT: the Contractor shall provide for all necessary watching and lighting and care of the whole works from weather, theft or other damage. All materials on site shall be protected from damage or loss.

STABILITY: the contractor shall accept responsibility for the stability and structural integrity of the works and support as necessary. Prevent overloading.

TRESPASS AND NUISANCE: all reasonable means shall be used to avoid inconveniencing adjoining owners and occupiers. No workman employed on the works shall be allowed to trespass upon adjoining properties. If the execution of the Works requires that workmen must enter upon adjoining property, the necessary permission shall be first obtained by the Contractor.

The Contractor shall not obstruct any public way or otherwise permit to be done anything which may amount to a nuisance or annoyance, and shall not interfere with any right of way or light to adjoining property.

TRAFFIC AND POLICE REGULATIONS: all traffic and police regulations particularly relating to unloading and loading of vehicles must be complied with and all permits properly obtained in due time for the works.

CONTROL OF NOISE: ensure that all measures are taken to control noise levels in accordance with the Noise at Work Regulations 1989, the control of Pollution Act 1974, the control of Noise (Code of Practice for Construction Sites) Order 1975 and BS 5228 including complying with DOE advisory leaflet 72 – noise control on building sites. Compressors, percussion tools and vehicles shall be fitted with effective silencers of a type recommended by the manufacturers of the compressors, tools or vehicles. The use of pneumatic drills and other noisy appliances must have the Contract Administrator's consent. The use of radio or other audio equipment will not be permitted.

POLLUTION: take all reasonable precautions to prevent pollution of the works and the general environment. If pollution occurs inform the appropriate Authorities and the Contract Administrator without delay and provide them with all relevant information.

NUISANCE: take all necessary precautions to prevent nuisance from smoke, dust, rubbish, vermin and other causes.

SAFETY HEALTH AND WELFARE: allow for complying with all Safety, Health and Welfare Regulations appertaining to all workpeople on site including those employed by sub-contractors and professional advisers, including but not limited to the following and to include any future amendments or re-enactments thereto:-

i) The Construction (Design and Management) Regulations 1994

ii) The Construction Regulations 1961, 1966 and 1996

iii)	The Factories Act 1961
iv)	The Offices, Shops and Railway Premises Act 1963
v)	Work Equipment Regulations 1998
vi)	Management of Health and Safety at Work Regulations 1999
vii)	The Health and Safety at Work, etc Act 1974
viii)	Manual Handling Operations Regulations 1992
ix)	Workplace (Health Safety and Welfare) Regulations 1992
x)	Personal Protective Equipment at Work Regulations 1992
xi)	Display Screen Equipment Regulations 1992
xii)	Special Waste Regulations 1996
xiii)	Control of Asbestos at Work Regulations 2002
xiv)	Construction (Head protection) Regulations 1989
xv)	Construction (Health and Safety and Welfare) Regulations 1996
xvi)	Control of Substances Hazardous to Health Regulations 2002
xvii)	Environmental Protection Act 1990
xviii)	Reporting of Injuries, Diseases and Dangerous Occurrences Regulations 1995
xix)	Waste Management Licensing Regulations 1994
xx)	Gas Safety (Installation and Use) Regulations 1994
xxi)	Fire Precautions (Workplace) Regulations 1997
xxii)	Lifting Operations and Lifting Equipment Regulations 1998
xxiii)	Provision and Use of Work Equipment Regulations 1998
xxiv)	The Secure Tenancies (Right to Repair) Regulations 1985
xxv)	The Equal Pay Act 1970
xxvi)	The Employment Protection (Consolidation) Act 1994
xxvii)	The Enterprise Act 2002
xxviii)	The Sex Discrimination Act 1975 and Disability and Discrimination Act
xxix)	The Race Relations Act 1976 and the Race Relations Amendment Act 2000
xxx)	Health and Safety (Safety, Signs and Signals) and [Consultation with Employees] Regulations 1996
xxxi)	Management of Health and Safety at Work and Fire Precautions Regulation 2003

A copy of the Contractor's Health and Safety Policy/ Health and Safety Plan as applicable shall be produced for inspection by the Contract Administrator/ Planning Supervisor.

The Contractor shall be responsible for ascertaining whether execution of any order for work complies with their requirements under the CDM Regulations, and must notify the Contract Administrator/CDM Co-ordinator of same and obtain approval prior to commencement of site activities

The Contractor will be responsible where applicable, for providing all information required by the Contract Administrator/ CDM Co-ordinator to update the Health and Safety File.

In the event of the Contractor ascertaining that execution of an order will or may involve interference with any hazardous substance or installation then the Contractor shall forthwith notify the same to the Contract Administrator and in so doing shall notify him in writing of any precautions proposed to be taken in consequence of the hazard which may affect the use of the premises or the comfort or freedom of movement of any person likely to be in or near the premises during execution of the order.

The Contractor shall likewise notify in writing the occupant of the premises, or the person in charge of the occupants or users of the premises on which works are in progress or about to be carried out, all restrictions guidance or other precautions which are desirable or necessary for the safety of all persons occupying or using the premises in consequence of the works. The Contractor shall provide all barriers and warning notices required for that purpose and shall make effective arrangements for the occupant or person in charge to consult and communicate with the Contractor, throughout the duration of the works, on the effects and nature of such precautions.

ASBESTOS SAMPLES: in the case of asbestos, the Contractor shall comply with the Control of Asbestos at Work Regulations 2002 and arrange for any necessary sampling and analysis before undertaking any work effecting the suspect material, and shall give the necessary notice to the Health and Safety Executive.

WORKING WITH ASBESTOS: when carrying out work on asbestos based materials, the Contractor shall comply with the Control of Asbestos at Work Regulations 2002 and appoint a licensed specialist sub-contractor. This requirement will be strictly enforced.

Any work with asbestos cement [eg. cleaning, painting, repair or removal]; any work with materials of bitumen, plastic, resins or rubber which contain asbestos, the thermal and acoustic properties of which are incidental to its main purpose and minor work with asbestos insulation, asbestos coating and asbestos insulating board which because of its limited extent and duration does not require a licence (eg. drilling holes, repairing minor damage, painting, removal of angle panel etc.) shall be carried out in accordance with the Approved Code of Practice and Guidance entitled Work with Asbestos which does not normally require a licence (Fourth edition) published by the Health and Safety Executive. The rates for such work shall be agreed with the Contract Administrator.

EMPLOYER'S SAFETY POLICIES: without prejudice to the Contractor's general obligations to ensure compliance with all statutory requirements relating to health and safety, the Contractor shall in particular observe and comply with:

(a) any specific condition, warning or direction given by the Contract Administrator on any matter relating to health and safety;

(b) the relevant provisions of any Employer's Safety Policy applicable to operations of the type in question when undertaken by Employer's employees, being a Safety Policy of which a copy has been given to the Contractor at or before the start of the work:

and

(c) any method statement agreed with the Contractor before the work begins identifying the safety precautions to be taken.

FIRE PRECAUTIONS: the contractor shall take all necessary precautions to prevent personal injury, death and damage to the works or other property from fire. Comply with Joint Code of Practice 'Fire Precaution on Construction Sites' published by the Construction Confederation and the Fire Protection Agency. Smoking will not be permitted on the works.

CANCELLATION ON DEFAULT: in the event of default by the Contractor in the proper observance of any necessary health and safety requirements, cancellation of the written order by the Contract Administrator shall not result in the Employer being obliged to reimburse either any costs incurred by the Contractor or the value of any abortive work except to such extent (if any) as those costs or that abortive work were incurred or performed without contravention of the health and safety requirement in question.

MAINTENANCE OF PUBLIC ROADS: the Contractor shall make good all damage to public roads, kerbs and footpaths, lawns etc, occasioned by exceptional traffic, delivery of materials and building operations generally to the reasonable satisfaction of the Contract Administrator and the local authority.

EXISTING MAINS AND SERVICES: the Contractor shall maintain during the progress of the works, the existing drainage system, water, gas, sewers, electric and other services and is to make arrangements for their continuance and take all necessary steps to protect and prevent damage to them. Should any mains, services ducts or lines be found to be in the way of new works, or require any attention, the Contractor is to seek instructions from the Contract Administrator.

Where it is necessary to interrupt any mains or services for the purpose of making either temporary or permanent connections or disconnections, prior written permission shall be obtained from the Contract Administrator and where appropriate from the local authority or public undertaking and the duration of any interruption kept to a minimum.

PROTECTION, DRYING AND CLEANING: the Contractor shall undertake the following, the cost of which shall be included in the tendered percentage adjustment to the prices in the schedule of rates:-

1. protect all work and materials on site, including that of Sub-Contractors, during frosty or inclement weather.

2. protect all parts of existing buildings which are to remain using polythene/dust sheets and make good any damage caused.

3. prevent damage to existing furniture, fittings and equipment left in the property. Cover and protect as necessary.

4. protect the adjoining properties by screens, hoardings or any other means to prevent damage or nuisance caused by the Works.

5. dry out the Works as necessary to facilitate the progress and satisfactory completion of the Works. The permanent heating installation may be used for drying out the works and controlling temperature and humidity levels, but:
    a) the Employer does not undertake it will be available
    b) the Contractor must take responsibility for operation maintenance and remedial work, and arrange supervision by and indemnification of the appropriate subcontractors and pay costs arising.

6. protect and preserve all trees and shrubs except those to be removed.

7. treat or replace any trees or shrubs damaged or removed without approval.

7. clean the Works thoroughly removing all splashes, deposits, rubbish and surplus materials.

DAMAGE: the Contractor shall exercise great care at all times to prevent damage to the building structure, fittings, furniture, equipment, finishes or the like and shall make good any damage caused by him at his own expense. In this connection all carpets, desks and other furniture or equipment including telephones in the vicinity of the work shall be covered by the Contractor with protective dust sheets or the like prior to any work commencing. Where it is necessary to use any naked flame or welding equipment in executing the work and where combustible materials are in use, adequate protection shall be given to other adjacent materials and personnel. Suitable fire extinguishers shall be provided and readily available at the position where such work is proceeding. The Contractor shall maintain the designated escape routes and exit doors within any building clear of all materials and plant at all times. The Contractor shall consult the Premises Manager at any of the locations in respect of precautions which should be taken for the safety of other occupants prior to the commencement and whilst work is in progress.

PERMIT TO WORK: before commencing any portion of the work the Contractor must establish the need for and if necessary obtain a Permit to Work.

The Contractor's tendered percentage adjustment to the Schedules of Rates shall be deemed to include allowances for time spent in obtaining permits for each portion of the work and claims made by the Contractor shall not be entertained by the Employer in respect of time lost in connection with the issue of permits. However the nett cost of fees and charges by local authorities is reimbursible to the Contractor by the Employer.

Permits to Work will be required for, but not limited to:-

- a) Excavation Work
- b) Hot Work/ Welding
- c) Confined space entry
- d) Cutting through or disconnecting existing services
- e) Access to roofs

Specific limitations on method/sequence/timing

SITE VISITS: before tendering the Contractor shall visit the site(s) and ascertain:-

1. Local conditions.

2. Means of access to the site(s).

3. The confines of the site(s).

4. Restrictions in respect of loading and unloading vehicles.

5. Factors affecting the order of execution of the work and the time required for the execution of the works.

6. The supply of and general conditions affecting labour, materials and plant required for the execution of the work.

POSSESSION OF THE SITE: no restriction. *

OR

POSSESSION OF THE SITE: possession of the site by the Contractor will be restricted as follows:-

WORKING SPACE : working space is limited to .... *

WORKING SPACE : take reasonable precautions to prevent workmen, including those employed by Sub-Contractors from trespassing on adjoining owner's property and any part of the premises which are not affected by the Works.

* Contract Administrator to complete/delete as applicable

WORKING HOURS: working hours are limited to the normal working hours defined in National Working Rules for the Building Industry as appropriate to the area in which the Works are located. Overtime shall not be worked by operatives on the site without the prior express permission in writing from the Contract Administrator.

OCCUPIED PREMISES: where work is done in occupied premises the Contractor shall take all reasonable care to avoid damaging the property or contents and shall make good all damage which arises from his work.

PROGRAMME OF WORKS: the Contractor shall prepare and submit for the approval of the Contract Administrator, a programme covering all aspects of the works.

SEQUENCE OF WORK OR OTHER RESTRICTION: .... [Note: any restriction on the work (eg. sequence or timing) must be given]. *

USE OF SITE: the site is not to be used for any purpose other than the execution of the contract.

REINSTATE SITE: confine to as small an area as practicable any operations which may affect the surface of the site and reinstate the site after the works are completed.

APPROVAL TO SITING: notify the Contract Administrator of the proposed siting of materials, temporary buildings, rubbish deposits and the like.

TIPPING: no allowance for tipping charges and Landfill tax charges in connection with materials obtained from site including those arising from demolition or alteration works has been included in the schedule of rates and costs should be included in the Contractors tendered percentage 'A'.

OVERTIME: where overtime is ordered in writing by the Contract Administrator, the Contractor will be paid the net additional cost, subject to the addition of 10% for overhead costs provided the contractor's returns are certified each week in relation thereto by the Contract Administrator. ✿

OVERTIME, NIGHT WORK AND INCENTIVES: all costs of overtime or night work at the discretion of the Contractor must be borne by the Contractor and no claims for additional payment in this respect will be allowed.

DAYWORKS: no work will be allowed as daywork unless previously authorised by the Contract Administrator and confirmed in writing. All vouchers specifying the time daily spent upon the work (and, if required by the Contract Administrator, the workmen's names) and the materials used properly priced and extended, shall be signed by the Contract Administrator.

Where daywork is authorised, the Contract Administrator shall be notified of the commencement and completion of the work, and the items of plant and workpeople concerned are to be solely engaged thereon and not employed upon any other work during progress of the daywork.

BUILDING OPERATIONS IN WINTER: the Contractor must be conversant with the measures and operations described in the booklet 'Winter Building' published on behalf of the DOE and obtainable from HMSO for ensuring continuity of work and productivity during inclement weather. The operations and measures described in the booklet shall be taken wherever practicable and having regard to nature, scope and programme of the Works.

✿ This percentage will only apply if no other percentage is inserted in appendix item 12 of the JCT MTC Conditions or is otherwise included in the Contract.

* Contract Administrator to complete/delete as applicable

Facilities/Temporary work/Services

NOTICES AND FEES TO LOCAL AUTHORITIES AND PUBLIC UNDERTAKING: such fees, charges, rates and taxes paid by the Contractor shall be reimbursed nett to him by the Employer. (See also WORKS BY PUBLIC BODIES and PERMIT TO WORK)

WORKS BY PUBLIC BODIES: the .... (name of * local authority or public undertaking) will carry out the following work which will be covered by provisional sums as follows:

Description of Work	Provisional Sum
....	....
....	.... *

The Contractor is to allow for general attendance and overheads.

The .... (name of local authority or public undertaking) will be carrying out, in accordance with their statutory obligations, the following work which is not part of the Contract Works: *

Description of Work .... *

[NB: this description should include the scope and timing of the work and its effect on the Contractor's operations]

OFFICES FOR PERSON-IN-CHARGE AND FACILITIES FOR EMPLOYER: the Contractor is to provide, erect and maintain suitably equipped offices for the Person-in-Charge and other necessary staff including Clerk of Works and provide heating and lighting and attendance throughout the duration of the Contract and remove and clear away upon completion and make good all work disturbed.

SITING OF TEMPORARY BUILDINGS ETC: all offices, messrooms, storage sheds, sanitary accommodation and temporary buildings shall be sited to the approval of the Contract Administrator. All areas so used must be made good on completion.

SANITARY ACCOMMODATION: the Contractor is to provide adequate suitable and proper sanitary accommodation which must be water-borne to the satisfaction of the Authority's Health Department and washing facilities for workmen to the standard required by the current Working Rule Agreement and to the satisfaction of the Health Department of the local authority and keep same in a clean and sanitary condition and remove and make good upon completion.

WORKING PLATFORMS: are to be provided to enable the work to be safely and effectively carried out. In all cases the Contractor shall comply with the requirements of the Construction (Health and Safety and welfare) Regulations 1996 together with other relevant regulations and requirements, consulting the Contract Administrator where differing provisions of scaffolding are possible. Agreement by the Contract Administrator to a particular method of scaffolding shall not relieve the Contractor of his responsibility for fully complying with Health and Safety at Work provisions. The Contractor's percentage addition to the prices in the Schedule of Rates is to include for:-

1. temporary staging to provide a working platform up to a height of 1.5 metres.

2. ladders for access to buildings of not more that two storeys in height.

- Contract Administrator to complete/delete as applicable

SCAFFOLDING: scaffolding or towers or mobile towers to provide working platforms greater than 1.5 metres in height will be dealt with as follows:-

1. by the use of the rates for scaffolding contained in Part 2 (Schedule of Descriptions and Prices).

2. by agreement between the Contract Administrator and Contractor.

3. by daywork.

4. by quotations from not less than three specialist firms tendering in competition.

SCAFFOLDING must be constructed in accordance with the requirements of the Health and Safety at Work Act 1974, the Management of Health and Safety at Work Regulations 1992 etc and subsequent amendments or re-enactments thereto and comply with:-

1. BS 5973: 1981 "Access and Working Scaffolds and Special Scaffold Structures in Steel".
2. BS 5974: 1982 Code of Practice for Temporary Installed Scaffold and Access Equipment.

SHORING, SCREENS, FENCING AND HOARDINGS: will be dealt with as follows:-

1. by the use of the rates for such non-mechanical plant contained in Part 2 (Schedule of Descriptions and Prices).

2. by agreement between Contract Administrator and Contractor.

3. by daywork.

4. by quotations from not less than three specialist firms tendering in competition.

NOTICE BOARD: upon written application the Contractor may display and maintain in an approved position a notice board stating his name and that of authorised Sub-Contractors.

PROVISION OF SKIPS: application must be made to the appropriate local authority department for the siting of any skips required for the collection and removal of contractors' waste and rubbish. No allowance for charges in connection with the use of skips has been included in the schedule of rates and costs should be included in the Contractor's tendered percentage. The cost of removal of rubbish on site not arising from the contract works such as fly tipping or spoil from occupants or other contractors is reimbursable.

The Contractor is to ensure that his tender provides for removing rubbish from the site both as it accumulates from time to time and at completion of the works.

Ensure that non-hazardous material is disposed of at a tip approved by a Waste Regulation Authority. Remove all surplus hazardous materials and their containers regularly for disposal off site in a safe and competent manner as approved by a waste regulation authority and in accordance with relevant regulations. Retain waste transfer documentation.

TEMPORARY TELEPHONE: the Contractor is to allow for providing temporary telephone facilities to the site and defray all charges in connection therewith, including the costs of all calls made by his own employees and those of any Sub-Contractors. No provision need be made for telephones for the Employer's representatives.

WATER FOR THE WORKS: the Contractor shall provide all water required for use in the works, by him or by his Sub-Contractors, together with any temporary plumbing, standpipes, storage tanks and the like, and remove on completion. He shall pay all fees and charges in connection therewith and make good all work disturbed.

LIGHTING AND POWER FOR THE WORKS: the Contractor shall provide all artificial lighting and power (electricity and/or gas) for the works, including that required by Sub-Contractors, together with any temporary wiring, switchboards, distribution boards, poles, brackets, etc. and remove same on completion, and pay all fees and charges in connection therewith and make good all work disturbed.

*Note: Where miscellaneous or improvement works to services installations is carried out under a subcontract the services contractor should identify the extent and nature of temporary lighting and power which will be provided by the general contractor at no charge.

METER READINGS: where charges for service supplies need to be apportioned ensure that meter readings are taken by relevant authority at possession and/ or completion as appropriate. Ensure that copies of readings are supplied to interested parties.

Operation/Maintenance of the finished works

Where appropriate the Contractor will provide the Contract Administrator with a free copy of the manufacturers' maintenance/operation manuals for installed equipment.

Submit a copy of each test certificate to the Contract Administrator as soon as practical and keep copies of all certificates.

CONTRACTOR'S GENERAL COST ITEMS

The contractor is to allow in his tendered percentage for the cost of the following:

MANAGEMENT AND STAFF including all overheads, offices, equipment, insurances, travel and expenses, supervision, programming, quantity surveying support staff and the like.

SITE ACCOMMODATION including erection, dismantling, hire charges, maintenance, services, charges, insurances etc. for offices, stores, canteens, compounds, sanitary facilities and the like.

SERVICES AND FACILITIES where not provided by the Employer at no charge to the Contractor, including power, lighting, fuels, water, telephone, security and the like.

MECHANICAL PLANT where not shown as included in Part 2 Schedule of Descriptions and Prices or provided by the Employer at no charge to the Contractor, including cranes, hoists, transport and other mechanical plant.

TEMPORARY WORKS including temporary access roads, hardstandings, hoardings, fans, fences and the like. (The Contract Administrator should define the exact requirements at tender stage)

TRAVELLING AND TRANSPORT OF LABOUR, PLANT AND MATERIALS within the boundaries specified as the 'Contract Area' to the properties listed within the Appendix item 1a including all expenses and vehicle costs.

CLIENT TRAINING: explain and demonstrate to the Employer's premises staff the purpose, function and operation of the installations including all items and procedures listed in the Building Manual.

SPECIFICATION REQUIREMENTS: any adjustments required to the schedule to comply with the specification, drawings or site requirements.

RECORDS AND DRAWINGS of the work as required by the Contract Administrator

GENERAL BONDING AND EARTHING as required by the IEE Regulations for any electrical work

*Contract Administrator to delete/ complete as applicable

IDENTIFICATION AND LABELLING of services specified

BUILDERS WORK INFORMATION, including marking up of any holes or chases, and provision of any sleeving requirements for pipes, conduits, ducts and the like

PAINTING AND PRIMING of services provided

WASTE as generated by the installations including cutting of materials to suit, loss, damage and the like

WORKING in complete conjunction with any ceiling installations

CLEANING of the services works and making good as required

INSPECTION TESTING AND COMMISSIONING of the works. Where testing, balancing and commissioning exceeds the scope of works, the Contract Administrator is to notify the Contractor of his requirements, and an hourly charge or lump sum agreed with the Contractor, if required.

WORK/MATERIALS BY THE EMPLOYER

Plant or materials may be provided by the Employer under the provision of contract clauses 1.6 to 1.12 inclusive.

NOMINATED SUB CONTRACTORS AND SUPPLIERS

Although the JCT Measured Term Contract does not provide for them the Employer may wish to nominate specialist sub-contractors and/or suppliers to carry out works and or supply goods which are not included in the schedule of descriptions and prices. The Contractor will be entitled to 2½% cash discount on the nett cost of the specialist work and 5% on the nett cost of specialist goods after deduction of trade and other discounts, and will also be allowed a percentage addition for profit, attendance and all overheads on Sub-Contractor's accounts after the adjustment to provide the cash discounts described above.**

Attendance includes general attendance and unloading, storing and placing goods in position for fixing, returning crates and packing etc.

(Prices are included for fixing only items which may have been ordered from a nominated supplier or supplied by the Employer). Separate provision shall be made for special attendance in accordance with the Standard Method of Measurement.

WORK BY STATUTORY AUTHORITIES

Where the Contract Administrator orders the Contractor to instruct Statutory Authorities to carry out works under the contract, the Contractor is entitled to recover the full cost of any fees and charges payable in consequence thereof and will be allowed 10% on the net charge to be added for profit, attendances and all overheads.

** This percentage does not apply to Sub Contractors work ordered under the daywork provision in Appendix 11 of JCT MTC conditions unless percentage adjustment for this item has not been included.

Appendix A

List of supplementary clauses which may be applicable on a particular contract.

1. Abortive Calls

2. Administration: Call out procedures
   Reports
   Evaluation
   Complaints procedure, penalties, damages, right of appeal

3. Certificate of Non-collusion

4. Contracting Associations (NICEIC, CORGI and HVCA)

5. Provision of a Contract Bond

6. Communications

7. Data Protection Act

8. Emergency services

9. Employer's obligations and restrictions

10. Good practice and examples

11. Identification

12. Inspections

13. Tenderers proposed method of working

14. Materials, goods and workmanship

15. Notice to occupiers

16. Obligations for statutory tax deduction scheme

17. Annual performance review

18. Potential hazards

19. Protection of furniture

20. Protection of gardens

21. Rehabilitation of Offenders' Act 1974

22. Definition of repairs : Day to day responsive repairs
    Package maintenance works
    Void property
    Out of hours emergency work

23. Response times

24. Security arrangements

25. Safety of children

Appendix A (Continued)

26. Spot audits
27. Smoking policy
28. Tests
29. Work equipment
30. Use of chlorofluorocarbons

Appendix B

1.0 Transfer of Undertakings and Protection of Employment Regulations

1.1 The present contract is being performed by an outside contractor. The Employer has no view as to whether or not the European Acquired Rights Directive No 77/187 and/or the Transfer of Undertakings (Protection of Employment) Regulations 1981 ("TUPE") applies to this Contract. It is up to each tenderer to form his own view on this. The Employer proposes to ask the present contractor whether he is of the opinion that TUPE might apply to this Contract and if so to provide a list of posts and details relating to it which he anticipates might transfer should TUPE apply. The Employer does not accept any responsibility for whether this information is made available, or if it is correct or not. The Tenderer may be required to complete a confidentiality agreement in respect of this information before it is made available to him.

1.2 The Tenderer must indicate whether his tender is based on TUPE applying or not applying. If the Tenderer has indicated that his tender is based on TUPE applying, he will be taken to have accepted that he will accept a transfer of any staff that should transfer to his contract. If the Tenderer's tender is successful, he must take any issues that may arise about transferring staff (including any questions as to who might transfer to his contract) directly with the present contractor. The Employer will not be willing to become involved in this.

1.3 If TUPE applies the Tenderer should take into account the following requirements in respect of transferring staff,

   1. The need to consult with recognised trade unions.
   2. The need to maintain existing rates of pay and conditions of employment.
   3. The need to provide pension arrangements broadly comparable to those provided at present to transferring employees (optional requirement).
   4. That liability will transfer to the successful contractor for any claims by transferring employees for redundancy, unfair dismissal or arising out of their previous employment even before transfer.

   It will be the Tenderer's responsibility, not the Employer's, to do this.

1.4 The Tenderer is expected in pricing his tender to make his own allowances for and accept the risk of fluctuations in his staffing availability or requirements. The Employer will not accept any tender in which the Tenderer's pricing varies according either to the number, identity or pension status of the staff he requires to perform the Contract or to any changes in the Tenderer's wage rates except so far as they may be reflected directly or indirectly in any method provided in the Contract Conditions for an annual review of the Tenderer's prices.

1.5 If the Tenderer has indicated that his tender is based on TUPE not applying, he must submit with his tender a written statement explaining why he believes TUPE will not apply if his tender is successful. Although this is primarily a matter between the outgoing and incoming Contractor, the Employer is concerned that the transition between contracts should be as seamless as possible. In the interests of continuity if the Employer does not agree with the Tenderer's contention that TUPE does not apply it reserves the right either to reject the tender or to evaluate it according to the Employers own assessment of the financial and other implications based on the Employers' own views as to the applicability to TUPE.

1.6 If the Employer accepts a bid on the basis that TUPE does not apply, it will require the successful tenderer in writing.

   1. To accept the risk that TUPE and/or the Directive might apply.
   2. To accept that if either is held to apply it will indemnify the Council against any cost that may fall on the Council in respect of any claims under TUPE or the Directive.
   3. To agree that if TUPE or the Directive are held to apply, the tenderer will not seek to rescind, repudite, terminate or amend the Contract.

Appendix B (Continued)

2.0 TUPE and the expiry of this contract

2.1 The Employer cannot and does not propose to commit itself as to,

1. What will be its Service requirements after this contract has expired.
2. What arrangements it may propose to make to procure the Service, or
3. What the legislative regime will be at that time either as to procurement of services or transfer of staff.

2.2 It therefore will not enter into any commitment as to what might happen to the successful tenderer's staff at the expiry of the Contract.

SECTION G

MATERIALS AND WORKMANSHIP PREAMBLES

GENERALLY

Unless otherwise stated or contradicted Materials and Workmanship Preambles are to apply reciprocally between Work Groups.

Unless otherwise stated or contradicted the rates contained in Part 2 (schedule of Descriptions and Prices) are to apply reciprocally between Work Groups/ Sections.

The preambles contained in this section are for guidance only, to indicate the basis on which level of pricing has been made. Proprietary brand names have been specified in certain instances. Equivalents (other equal or approved) may be used if approved, in writing, by the Contract Administrator.

For the purpose of these preambles the words "Contractor" or "Sub-Contractor" both mean the person(s) carrying out the works to services installations as a Contractor or Sub-Contractor.

MATERIALS

Where and to the extent that materials are not fully specified they are to be suitable for the purposes of the Works stated in or reasonably to be inferred from the Contract, in accordance with good practice and complying with current British Standards and the recommendations contained within the current edition of the CIBSE guide.

Proprietary materials are to be handled and stored strictly in accordance with manufacturer's instructions and recommendations. Such materials are to be obtained direct from the manufacturer's or through their accredited distributors.

Where appropriate, items in the Preambles in other Sections shall apply equally to this Section.

Any preambles included within a specification issued with this document are to apply.

WORKMANSHIP

Where and to the extent that workmanship is not fully specified it is to be suitable for the purposes of the Works stated in or reasonably to be inferred from the Contract, in accordance with good practice and complying with current British Standards and the recommendations contained within the current edition of the CIBSE guide and current IEE regulations. Workmanship is to be of a high standard throughout.

Work liable to damage by frost is not to be carried out at temperatures less than 5 degrees Celsius unless precautions are taken against low temperatures. Submit details of such precautions to the Contract Administrator.

Ensure that site staff responsible for supervision and control of emergency, repair works and alteration/ small works are experienced in this type of work.

Items are to include for adequate temporary protection of building structures, decorations, furnishings, equipment and contents from damage by water, dust, debris and the like. Services connections for appliances which are to be removed or relocated are to be cut back and stopped off out of sight wherever possible and left in a neat arrangement.

Naked Flame: where it is necessary to use any naked flame or welding equipment in executing the Work and where combustible materials are in use adequate protection shall be given to other adjacent materials and personnel. Suitable fire extinguishers shall be provided and made readily available at the position where such work is proceeding. Designated escape routes and exit doors, etc shall be maintained and kept clear of all plant and materials at all times.

All reasonable fire precautions shall be taken in respect of stores workshops and other areas/ installations.

Stability of the works: no cutting through floors or walls or under foundations will be permitted other than that required by the drawings or schedules without the sanction of the Contract Administrator. Do not permit anything to be done which may injure the stability of the works or the existing services.

Draining and filling existing systems: where connections are made to an existing installation allowance shall be made for emptying, refilling and venting the existing and new installations during normal working hours.

Allowance shall be made for disconnecting and removing all redundant materials and equipment from site. Such materials shall become the possession of the Contractor and appropriate credit shall be allowed.

Unfinished works shall be left in a safe condition and suitably protected to prevent unauthorised access and interference.

Heavy equipment such as radiators, etc shall be stored in such a manner as to prevent falling or slipping and shall be protected to prevent unauthorised access and interference.

Statutory Authorities where the scope of works affects incoming services, arrange with appropriate Authority to carry out works as necessary to enable the Contract to be carried out.

Burning on site of materials arising from the work will not be permitted without prior approval.

Pipe sizes unless otherwise stated, tubes and their fittings are classified by their internal diameters.

Protection from any variety of damage whatsoever is to be included.

Duct covers, trenches, etc shall be replaced on leaving the site or the immediate vicinity or shall be adequately protected to prevent accidents.

In occupied premises, all works shall be arranged to minimise inconvenience to the normal running of the premises. Any interruptions to the existing services shall take place only with the prior approval of the person-in-charge.

Where thermal insulation to pipework, boilers, calorifiers, etc is suspected of containing asbestos, no works shall be carried out. The Contract Administrator shall be advised of the suspected asbestos and will issue further instructions as considered appropriate.

Workmanship preambles included in a specification issued with this document are to be included.

Delivery periods: immediately upon acceptance the Contractor shall verify the delivery periods of all materials required to complete the works, and must notify the Contract Administrator of any material delivery period which may have an adverse effect on progress.

All investigating work over and above that normally required for installation testing, commissioning, British Standards and Codes of Practice is to be agreed with the Contract Administrator as such prior to execution, and charges on a daywork basis or by a method previously agreed.

FIXINGS TO BUILDING FABRIC: the following shall apply

a) Preparation: mark-out, set out and firmly fix all equipment, components and necessary brackets and supports.

b) Manufacturer's drawings: use manufacturer's drawings and templates for purposes of marking and setting out.

c) Size of fixing: use largest size of bolt, screw or other fixing permitted by diameter of hole in item to be fixed.

d) Greasing of fixings: where indicated, ensure all bolts, screws or other fixings used are greased or suitably lubricated in accordance with manufacturer's instructions.

e) Standards: comply with BS 3974 Part 1 for fixings. Ensure that fixings such as expanding anchors are tested for tensile loading with BS 5080.

f) Plugs: use plugs of suitable size and length for fixings. Use plastic, fibrous or soft metal non-deteriorating plugs to suit application. Do not use wood plugs.

g) Screws: use screws to BS 1210. Generally use sheradized steel wood screws for fixing to concrete, brickwork or blockwork. Grease screws where indicated. In damp or exposed situations use greased brass wood screws.

h) Shot fired fixings: obtain approval prior to use of shot fired type fixings.

i) Self adhesive fixings: use self adhesive type fixings where indicated.

j) Drilling: drill holes vertical to work surface. Use drills of requisite size and depth, and appropriate to fabric. Flame-cut holes in metal work are not permitted.

k) Fixing to reinforced concrete: take precautions to avoid fixing through reinforcement.

l) Fixing to brickwork: do not fix to unsound material or mortar between brickwork courses.

MECHANICAL SERVICES

MATERIALS

PIPEWORK:

MDPE	:	comply with BS 6572, with compression or fusion fittings to BS 864 Part 3
UPVC	:	comply with BS 3505 : 1986 : solvent welded fittings to BS 4346 Part 1 : 1969
Copper	:	comply with EN1057 R250 (TX) for all services where specified above ground, table Y where buried. Compression or capillary fittings to BS 864 : Part 2
Carbon Steel	:	comply with BS 1387 : 1985 –

    1. Medium grade or as specified

    2. Plain ends or screwed ends

    3. Finish as specified

    4. Screwed fittings comply with BS 143 and BS 1256

    5. Butt welded fittings comply with BS 1965 : Part 1

PIPEWORK JOINTING:

Flanges   :   comply with BS EN 1092 Parts 1 to 4

    1. Material as specified

    2. Flange type as specified

    3. Jointing rings as specified

Screwed Joints   :   comply with BS 21

    1. Hemp and jointing compound comply with BS 5292

    2. PTFE tape comply with BS 7786

Welded joints

    1. Steel: gas welding BS 1453 type A2 or A3

    2. Steel: electric arc welding BS 2633

    3. Copper: bronze welding BS 1453

Unions

    1. Bronze to bronze navy pattern

    2. Bronze to iron railroad pattern

PIPEWORK ANCILLARIES:

Stop valves	:	gate type: comply with BS 5154 incorporating copper alloy body
	1.	Ends as specified
Stop valves	:	gate type: comply with BS EN 1171 incorporating cast iron body with flanges to BS EN 1092 Parts 1 to 4
	1.	Seats and discs as specified (including variable orifice)
Double regulating valves	:	globe type: comply with BS 5154 incorporating copper alloy body
	1.	Pattern as specified
	2.	Cast iron body type: comply with BS EN 13789
Radiator valves	:	comply with BS 2767: wheel valves on flow connections and lockshield valves on return connections unless otherwise indicated
	1.	Pattern as specified
	2.	Ends as specified
	3.	Finish as specified
Check valves	:	swing check type: comply with BS 5154 incorporating copper alloy horizontal monitoring full bore type body
	1.	Ends as specified
	2.	Cast iron body type: comply with BS EN 12334: flanged to BS 4504
Direct acting	:	comply with BS 759 : safety valves
	1.	Locking device as specified
Drain cocks	:	throughway gland cock type
Strainers		
	1.	Material as specified
	2.	Ends as specified
	3.	Pattern as specified
Gauges	:	comply with BS 89, of robust construction, enclosed in dust tight metal cases. Retained dial glass with bezels screwed to case. Finish with chromium plating
	1.	Dial case size and type as specified
	2.	Dial scale type as specified
	3.	Mounting requirements as specified

Temperature Gauges	:	suitable for use with pocket and provided with gland attachment on thermometer stem
	1.	Types as specified
	2.	Pockets as specified
	3.	Secondary elements as specified
	4.	Dial case and size and type and scale type as specified
Pressure and altitude gauges	:	vapour pressure type: comply with BS EN 837 Parts 1 and 2. Connect to pipeline systems via matched gauge cocks and cock connectors
	1.	Dial case size and type, and scale type as specified

PUMPS:

Heating and hot water	:	shall be of the glandless, in-line, single stage centrifugal type, with high torque motors, which are easily removable

INDIRECT CYLINDERS:

Shall be horizontal or vertical type pattern, of galvanised mild steel construction: comply with BS 1565, or copper construction: comply with BS 1566 or BS 853 Part 2

1. Supports as specified
2. Connectors and ancillaries as specified
3. Combination types as specified
4. Duty and temperatures as specified

THERMOSTATIC MIXING VALVES:

Shall be galvanised or brass construction with chromium plated finish. Each valve shall be complete with a back plate for wall mounting unless otherwise stated

1. Valve operation as specified
2. Operating temperature as specified

THERMOSTATIC RADIATOR VALVES:

Shall be fixed head or remote temperature adjuster type as specified, with copper alloy body and angle pattern with spherical seated disc

1. Connection as specified
2. Finish as specified

WATER STORAGE TANKS:

Shall be polypropylene and comply with BS 4213 or galvanised: BS 417

1. Sizes and construction as specified
2. Ladders to comply with BS 4211
3. Supports to comply with BS 449

RADIATORS:

Shall be complete with all necessary water tappings for all valves and shall also include an air vent and necessary tapping.

1. Material and construction as specified

2. Mounting requirements as specified

PIPE COILS:

Shall include a valve on both flow and return connection, with reducing sockets and air vents as necessary.

1. Material and construction as specified

2. Mounting requirements as specified

CONVECTORS:

Heater battery installations as for pipe coils.

1. Material and construction as specified

2. Mounting requirements as specified

3. Controls as specified

UNIT HEATERS:

As for radiators and convectors.

RADIANT PANELS:

As for unit heaters.

THERMAL INSULATION:

Comply in general with BS 5422 and the British Standards publications referred to therein. Employ materials that comply with BS 476 Part 4, non-combustibility test, or obtain a class 'O' fire rating to Building Regulations, Section E15, when test to BS 426 Part 6 & 7. Thermal conductivity comply with BS EN 12664, BS EN 12667, BS EN 12939 and BS 2972.

1. Material and density as specified

2. Facings as specified

3. Aluminium sheeting to BS 5970 Part 5 or as specified

4. Reinforcement as specified

5. Thickness as specified

BOILERS:

Shall be complete with fully automatic burners and control systems.

    1. Equipment and its location for boiler shall be as the manufacturers requirements or as specified

    2. Dismantling, assembling, repairing and or testing shall be by the manufacturer or as specified

    3. Housing/ casing and commissioning requirements as above

BOILER MOUNTINGS:

Fit boiler mountings to comply with BS 759 and BS 779, Section 4; BS 855, Section 5 or BS 2790 as appropriate.

BOILERHOUSE INSTRUMENTATION AND CONTROLS:

Supply and fix instrumentation to enable user to check usage and efficiency of boiler plant. Ensure that boiler automatic controls supplied by boiler manufacturer comply with BS 779, Section 6 or BS 855, Section 7 as appropriate and :-

    1. British Gas recommendations

    2. Shell UK Oil recommendations

GAS BURNERS:

Supply gas burner matched to boiler. Provide gas burners manufactured in accordance with:-

    1. British Gas requirements – rated below 60kW

    2. BS 5885 – rated above 60kW

VENTILATION TO BOILER ROOMS:

Ensure that ventilation to boiler room complies with CIBSE Guide Section B13 and Local Authority recommendations. For equipment with rated output less than 60kW comply with BS 5440 Part 2 for air supply to gas burner. Ventilate boiler room by natural or mechanical means.

CHIMNEYS AND FLUES:

Provide flues and chimneys as indicated. Ensure chimney heights comply with Clean Air Act Memorandum. Ensure fan diluted flue systems fitted on gas fired installations, are in accordance with British Gas recommendations. For outputs less than 60kW comply with BS 5440 Part 1 for installations of flue. Comply with BS 5854.

1. Provide sheet metal flue pipes and accessories for gas boilers in accordance with BS 715

2. Provide factory made insulated chimneys for oil fired boilers in accordance with BS 4543, Part 3

3. Provide cast iron spigot and socket flue pipes and fittings in accordance with BS 41

4. Provide clay flue linings and flue terminals in accordance with BS 1181

5. Provide steel chimney in accordance with BS 4076

6. Install monolithic lining in steel chimneys in accordance with BS 4207

7. As specified

DUCTWORK:

Duct	:	comply with DW 142 Part 3 Sections 8 & 9 for construction of rectangular ductwork
Fittings	:	comply with DW 142 Part 3 Section 11 for manufacture of all fittings for rectangular ducts
Finish	:	zinc coated steel – hot dip galvanised

DUCTWORK ACCESSORIES AND COMPONENTS:

FLEXIBLE JOINTS: Manufacture and install flexible joints as detailed in DW 142 Part 7 Section 26 or DW 151 Section 11, as appropriate. Comply with BS 476 Parts 6, 7 and 8.

Position : use flexible joints

1. to make connections to air diffusers, grilles and air registers, where indicated

2. on inlet and outlet connections to all fans or fan and attenuator assemblies, where indicated

Properties : fit flexible joints:-

1. with acoustic properties, to prevent noise breakout

2. for expansion

SEALANTS GASKETS AND TAPES: for sealing materials and method of use comply with DW 142 Part 7 Section 27 or DW 151 Section 12 as appropriate.

PROTECTIVE FINISHES: comply with DW 142 Part 7 Section 28 for provision of protective finishes to ductwork. Make good welding damage as DW 142 Part 7, clause 28.3.2. Where indicated provide protective finish by:-

Painting – galvanised sheet ducts

SPLITTERS: construct in a similar material to main ductwork and attach to duct as described in DW 142 Part 3 Section 11.4.

TURNING VANES-LOW AND MEDIUM PRESSURE: PROVIDE TURNING VANES IN LOW AND MEDIUM PRESSURE DUCTWORK OF:-

Single skin pattern

Constructed from material similar to main ductwork. Fasten turning vanes as described in DW 142 Part 3 Section 11.5.

DUCTWORK ANCILLARIES:

ACCESS OPENINGS: provide access openings in accordance with DW 142 Part 7 Section 21, DW 151 Section 10 or DW 191 Section 8 as appropriate.

ACCESS AND INSPECTION COVERS: provide purpose made access and inspection covers in positions indicated in accordance with DW 142 Part 7 Section 21.2. Fit access cover with quick release catches.

ACCESS DOORS: provide purpose made hinged access doors where indicated in accordance with DW 142 Section 21.3. Fit access doors with latch style fastenings. Size access doors:-

In accordance with DW 142 Part 7 Section 21.2.1

HANGERS AND SUPPORTS: provide hangers and supports throughout in accordance with DW 142 Part 6, DW 151 Section 7 or DW 191 Section 7 as appropriate.

CONTROL DAMPERS: provide control dampers in accordance with DW 142 Part 7 Section 22 or DW 151 Section 8 as appropriate. Supply control damper types as specified.

MOTORIZED CONTROL DAMPERS: provide control dampers complete with motor, linkage and drive rod where indicated with supports for motors.

FIRE DAMPERS: for details of construction of fire dampers and installation frame comply with DW 142 Part 7 Section 23 or DW 151 Section 9 as appropriate. All fire dampers to comply with BS 476 Part 8. Provide fire dampers types as specified.

FIRE DAMPER FUSIBLE LINKS: where indicated, supply spare fusible links for fire dampers. Supply links to fuse temperature indicated.

SMOKE DAMPERS: provide smoke dampers time rated as indicated.

INSECT GUARDS: provide insect guards where indicated.

MECHANICAL SERVICES

WOKMANSHIP

PIPEWORK:

APPEARANCE: arrange all exposed pipe runs to present neat appearance, parallel with other pipe or service runs and building structure, subject to gradients for draining or venting. Ensure all vertical pipes are plumb or follow building line.

SPACING: space pipe runs in relation to one another, other services runs and building structure, allow for specified thickness of thermal insulation and ensure adequate space for access to pipe joints, etc.

The following are recommended as minimum clearances in spacing of pipe runs:-

GRADIENTS: install pipework with gradients to allow drainage and/ or air release, and to the falls where indicated.

AIR VENT REQUIREMENT: as specified.

DRAIN REQUIREMENTS: grade pipework to allow system to be drained.

EXPANSION AND CONTRACTION: arrange supports and fixings to accommodate pipe movement caused by the thermal changes, generally allow the flexure at changes in direction. Allow for movement at branch connections.

PIPE FITTINGS: use excentric type reductions and enlargements on horizontal pipe runs to allow draining and venting, concentric on vertical pipes, with easy transition and an included angle not exceeding 30 degrees. Do not use bushes, except at radiators and at fittings where required size is not of standard manufacture. Where required, use eccentric bushes to allow draining or venting; maximum aspect ratio not to exceed two pipe sizes; above this ratio use reducing fittings. Use square tees at venting and draining points. Square elbows are not acceptable. Use bends where practical.

FABRICATED FITTINGS – FERROUS: supply pipe material and end connections to the specification of the associated straight pipe runs

Pattern Technique   :   Bends, springs, offsets and branches
                        :   Pipe bore 50mm or less – machine cold bend
                        :   Pipe bore greater than 50mm – machine hot bend

Ensure that fabricated branch bends of welding saddles are to the fitting proportions in BS 1965, Part 1.

PIPES THROUGH WALLS AND FLOORS: enclose pipes passing through building elements, (walls, floors, partitions, etc.) concentrically within purpose made sleeves. Fit masking plates where visible pipes pass through building elements, including false ceilings of occupied rooms

PIPE SLEEVES: cut sleeves from material same as pipe insulation if insulation is carried through sleeve, to allow clearance. Do not use sleeves as pipe supports. Install sleeves flush with building finish. In areas where floors are washed down install with a 10mm protrusion above floor finish.

Pack annular space between pipe and sleeve with mineral wool or similar non-flammable and fire resistant material to form a fire/ smoke stop of required rating. Apply 12mm deep cold mastic seal at both ends within sleeve.

© NSR 2008 2009

WALL FLOOR AND CEILING MASKING PLATES:

Material	:	Copper alloy, chromium plated
Type	:	Heavy, split on the diameter, close fitting to the outside wall of the pipe
Fixing	:	Chrome raised head fixing screws

CONNECTIONS TO EQUIPMENT: make final connections to equipment in accordance with manufacturer's instructions and as indicated.

DISTRIBUTION HEADERS: terminate ends with a cap, a blank flange or as indicated.

TEMPORARY PLUGS, CAPS AND FLANGES: seal all open ends as installation proceeds by metal, plastic or wooden plugs, caps or blank flanges, to prevent ingress of foreign matter. In the event of such precautions not being taken, pipework adjacent to open ends to be stripped out by such a point or points that will demonstrate that fouling of bores has not occurred.

WELDING GENERALLY: use skilled craftsman in possession of a current Certificate of Competence appropriate to type and class of work, issued by an approved authority. Mark each weld to identify operative. Submit specimen welds, representative of joints and conditions of site welding, for each craftsman, test non-destructively, approximately 10% of buttweld joints and 5% of all other joints

1) Welded Joints Class 1 – Weld joints to BS 1821 and BS 2633 as appropriate. Carry out non-destructive testing on 10% or as indicated

2) Welded Joints Class 2 – Weld pipeline joints to BS 2640 and BS 2971 and to HVCA Code of Practice TR5/5, welding of carbon steel pipework, as appropriate

3) Examine welds radiographically, percentage as indicated

WELDED JOINTS, STEEL PIPES: preparation, making and sealing

: Oxy-acetylene welding, comply with BS 1821 or BS 2640 appropriate to system temperature and pressure

: Arc welding, comply with BS 2633 or BS 2971 appropriate to system temperature and pressure. Use arc welding process on piping greater than 100mm

FLANGED JOINTS, STEEL PIPES:

Welded flanges	:	Weld neck and bore of 'slip on' flange Butt Weld neck on welding neck flange
Screwed flanges	:	Apply jointing materials. Screw on flange and expanded tube into flange with roller expander where necessary
Preparation	:	Ensure that flange mating faces are parallel; flange peripheries are flush with each other and bolt holes are correctly aligned
Making and sealing	:	Insert jointing between flange mating faces. Pull up joint equally all round

SCREWED JOINTS, STEEL PIPES:

Preparation	;	Ensure that plain ends are cut square. Reamer out bore at plain ends. Screw plain ends, taper thread.
Making and sealing	:	Coat male pipe threads with jointing compound and hemp, or PTFE tape on small sizes. Immediately after applying coating, connect with female end of socket or fitting, and tighten ensuring that coating does not intrude into pipe. Leave joint clean.

DISSIMILAR METALS: take appropriate means to prevent galvanic action where dissimilar metals are connected together.

PIPE RINGS AND CLIPS: select type according to the application attention where pipes are subject to axial movement due to expansion or contraction.

Use pipe clips to comply with BS 3974 Part 1, take into account the pipe load, material and pipe/insulation surface temperature.

PIPE SUPPORTS: arrange supports and accessories for equipment, appliances or ancillary fitments in pipe runs, so that no undue strain is imposed upon pipes. Ensure that materials used for supports are compatible with pipeline materials.

SUPPORT SPACING: space supports as tables

PIPE BORE (mm)	MAXIMUM SUPPORT SPACING (m)					
	STEEL	PIPE	COPPER	PIPE	IRON	PIPE
Nominal up to	Horiz	Vert	Horiz	Vert	Horiz	Vert
15	1.8	2.4	1.2	1.8	-	-
20	2.4	3.0	1.4	2.1	-	-
25	2.4	3.0	1.8	2.4	-	-
32	2.7	3.0	2.4	3.0	-	-
40	3.0	3.6	2.4	3.0	-	-
50	3.0	3.6	2.7	3.0	1.8	1.8
65	3.7	4.6	3.0	3.6	-	-
80	3.7	4.6	3.0	3.6	2.7	2.7
100	3.7	4.6	3.0	3.6	2.7	2.7
125	3.7	5.4	3.0	3.6	-	-
150	4.5	5.4	3.6	4.2	3.7	3.7
200	5.0	6.0	-	-	3.7	3.7
250	5.0	6.0	-	-	4.5	5.4
300	6.1	10.0	-	-	8.0	10.0
350	10.0	12.0	-	-	-	-
400	10.5	12.6	-	-	-	-
450	11.0	13.2	-	-	-	-
500	12.0	14.4	-	-	-	-
600	14.0	16.8	-	-	-	-

PIPE BORE (mm)	UPVC PIPE Class O,B,C	PIPE Class D,E,6,7	PE Type 32	PIPE Type 50	GLASS	PIPE
Nominal up to						
10	-	0.6	0.3	0.45	-	-
15	-	0.6	0.4	0.6	-	-
20	-	0.65	0.4	0.6	-	-
25	-	0.75	0.4	0.6	-	-
32	-	0.8	0.45	0.7	-	-
40	-	0.8	0.45	0.7	0.9	1.7
50	1.1	1.2	0.55	0.85	1.2	1.7
65	1.2	1.4	0.55	0.85	-	-
80	1.4	1.5	0.6	0.9	1.2	1.7
100	1.5	1.7	0.7	1.1	1.2	1.7
125	1.7	1.9	-	-	-	-
150	1.8	2.1	-	1.3	1.2	1.7
175	2.0	2.3	-	-	-	-
200	2.1	2.5	-	-	-	-
225	2.3	2.7	-	-	-	-
250	2.4	2.9	-	-	-	-
300	2.6	3.1	-	-	-	-
350	2.9	3.4	-	-	-	-
400	3.1	3.7	-	-	-	-
450	3.4	3.7	-	-	-	-
above 450	3.7	3.7	-	-	-	-

Space vertical support intervals for plastics pipe at not greater than twice horizontal intervals tabulated. Where multiple pipe runs of differing bores are supported from a common point, use support spacing of pipe requiring closest spacing.

Spacings given for UPVC to BS 3505 are for 20°C ambient or working temperature. Reduce spacing between supports for temperatures above 20°C. Support continuously for temperatures 60°C and above.

ISOLATION AND REGULATION: provide valves, cocks and stop taps for isolation and/ or regulation where indicated, and on:-

: Mains to isolate major section of distribution

: The base of all risers and drops except in cases where one item of apparatus only is served which has its own local valve or stop tap

: Points of pipe connection of all items of apparatus and equipment except where the item could conveniently be isolated or regulated by valves provided for other adjacent items

: Draw-off fittings except where ranges of fittings are served by a common float, the isolator then being fitted with the float

: As indicated

MAINTENANCE AND RENEWAL: arrange pipework, valves, drains, air vents, demountable joints, supports, etc. for convenient routine maintenance and renewals. Provide all runs with regularly spaced pattern of demountable joints in the form of unions, flanges, etc. and also at items of equipment to facilitate disconnection.

Locate valves, drains, flanges etc. in groups.

PROTECTION OF UNDERGROUND PIPEWORK: protect where indicated against corrosion by the application of a compatible anti-corrosive, non-cracking, non-hardening waterproof sealing tape.

Apply, after cleaning pipework, by wrapping contrawise with two layers spirally around the pipe, ensuring a 50% minimum overlap.

PROTECTION OF BURIED PIPES: provide earth cover as follows:-

Water pipework - 900mm minimum; 1200 maximum where practicable
Fuel oil and gas - 500mm minimum;
Under roadways provide minimum cover of 900mm

CLEANING: remove cement and clean off all pipework and brackets.

STEEL PIPEWORK: remove scale, rust or temporary protective coating by chipping, wire brushing or use of approved solvents and paint with one coat of red oxide primer, as work proceeds.

STEELWORK: prepare supports, bearers and other uncovered steelwork and steel pipework. Where not exposed, paint with one coat zinc chromate/ red oxide primer. Where indicated galvanise after manufacture.

NON-FERROUS COMPONENTS: thoroughly clean and degrease.

PIPEWORK ANCILLARIES:

INSTALLATION: install pipeline ancillaries to comply with BS ISO 9000.

POSITIONING OF COMPONENTS: locate flow and pressure measurement valves to ensure manufacturer's recommended straight length of pipe upstream and downstream of valve is provided.

POSITIONING OF DOUBLE REGULATING VARIABLE ORIFICE VALVE: install double regulating variable orifice valve to ensure equivalent of 10 diameters of straight pipe upstream and 5 diameters downstream of double regulating valve.

POSITIONING OF REGULATING STATION: install regulating station to ensure equivalent of 10 diameters of straight pipe upstream of metering station and 5 diameters downstream of double regulating valve.

VENT COCKS: provide outlets of vent cocks with discharge pipes.

VALVE STUFFING BOXES: adjust glands of all stuffing boxes at normal plant operating temperature and pressure in accordance with manufacturers instructions. Ensure that valve action is not impaired by other tightening.

DISCHARGE CONNECTIONS: fit pipework connections, where indicated, to provide:-

: discharge connection to Safety and Relief valves terminating at a safe discharge point

: discharge connection to vent cocks terminating 150mm above floor level

: bleed connection from air bottles terminating with air cock or needle valve in a convenient position

: discharge pipe to automatic air vents terminating over a suitable gully or drain line in a visible location

EXPANSION DEVICES: where expansion and contraction cannot be accommodated by selected route, provide pipework loops, as indicated. Limit total stress set up in material of pipe wall, taking into account components due to internal pressure, tension and bending to less than 69 MPa for steel pipelines and less than 51.5 MPa for copper pipelines. Where location does not permit sufficient flexibility, provide proprietary devices, as indicated.

EXPANSION BELLOWS INSTALLATION: provide guides and apply cold draw to manufacturer's recommendations.

ARTICULATED COMPENSATORS INSTALLATION: install to manufacturer's recommendations.

ANGULAR COMPENSATORS INSTALLATION: install to manufacturer's recommendations.

HOUSE COMPENSATORS INSTALLATION: provide at connections to equipment to protect plant from imposed stresses caused by movement of connected pipework.

PUMPS:

GENERAL: comply with manufacturer's recommendations regarding application and installation of pumps. For in-line pumps ensure that motor is positioned in accordance with manufacturer's requirements.

PIPELINE CONNECTIONS: mount pumps independently from connecting pipework to ensure no load is transmitted from pipework to pump casing on pump suction and discharge.

DRAIN LINES FROM PACKED AND WATER-COOLED GLANDS: provide drain lines complete with tundish from glands to nearest drain. To facilitate cleaning use plugged tees instead of elbows.

MATERIALS : Light gauge copper capillary fittings with copper tundish

: Galvanised tube : BS 1387, screwed : BS 21 with galvanised tundish

MOUNTINGS: mount motors for belt drive pumps resiliently. Bolt pumps down to concrete base.

ALIGNMENT: align pump to prevent undue restraint and thrust on interconnecting pipework. Align drives to prevent undue wear and restraint on pump shaft. For belt drives, align pulleys and tension belts to prevent undue wear and out of balance forces.

ACCESS: locate pump within the system with adequate space around it for service and maintenance.

MAINTENANCE REQUIREMENTS FOR SEWAGE PUMPS: for ease of service and maintenance, install submersible sewage pumps on guide rails or with lifting cables. Fit pumps with automatic discharge connections, which locate on to permanent pipework at low level in chamber.

INDIRECT CYLINDERS: each cylinder shall be provided with the following unless otherwise specified:-

a) Safety valve

b) Altitude gauge

c) Dial thermometer

d) Integral male or female bosses to suit primary, secondary, cold feed and vent connection as detailed

e) Drain cock

STORAGE TANKS:

Where sectional tanks are specified, the contractor shall be responsible for the correct assembly and sealing of the tank and shall fill the tank to test for leakage prior to the installation of pipework. On completion of a satisfactory test, the contractor shall clean off all excess jointing material from between the tank sections.

Unless otherwise stated all tanks shall be complete with a close fitting loose cover manufactured from the same material as the tank.

Connections for inlet, outlet and overflow pipework shall be made strictly in accordance with the tank manufacturer's recommendations.

Where copper pipework is to be connected to a GRP tank, the use of heat to make capillary joints will not be permitted. Compression fittings will be accepted adjacent to tanks only.

Where tank connectors are utilised for pipe connections, nylon sealing washers, installed dry, shall be used. The use of oil based sealing compound will not be acceptable.

Unless otherwise stated the base of all GRP tanks shall be supported over the full area of the base of the tank.

A stop cock shall be supplied and fixed on the cold mains supply to the cistern adjacent to the ball valve connection.

Outflow connections shall be provided with gate valves with a handwheel type on cold down and hot water services and a lockshield type on boiler feeds. These valves shall be fitted as close to the tanks as possible whilst remaining accessible at all times.

The heating feed and expansion cistern shall be fitted with a long arm ball valve set so that not more than 100mm of water will be in the cistern under unheated conditions. The body of the valve shall be fixed above cistern overflow level. Ball valve type to be specified.

RADIATORS:

Cast iron radiators without feet shall be secured to the walls by means of the top stays and bottom brackets manufactured by the radiator manufacturer. All top stays to have brass dome nuts and the number of top and bottom brackets used shall comply with the manufacturer's recommendations.

Cast iron radiators with feet are to have the above type and number of top stays as recommended.

Steel radiators shall be secured to the wall by means of the screw on type brackets with 50mm (minimum) long coachbolts at the centres recommended by the manufacturer.

Where manufacturers supply separate top and bottom brackets, both shall be securely fixed to the wall.

Unless otherwise stated radiators shall be suitable for bottom opposite end connections and each radiator is to be supplied with a manual air release cock at the return connection end.

Before radiators are ordered the Contractor shall verify the dimensions of radiators by site measurement or with the Contract Administrator.

Certain radiators shall be suitable for TBOE and TBSE connections as shown on the schedules. Where BOE connections are used with cast iron radiators, a blind nipple shall be used fitted between the first and second section at the flow end.

All radiators unless otherwise stated shall be mounted a minimum of 150mm above finished floor level.

Where radiators are mounted under windows the top of the radiator shall not project above the sill.

The Contractor shall be required to work in conjunction with others eg, plasterers, painters etc, and is to include for taking down and refixing twice the whole of the radiators as and when required by other trades.

PIPE COILS:

Low level coils shall be fixed at a minimum of 80mm clear of the finished floor level with a minimum distance of 65mm clear between pipes where applicable.

CONVECTORS:

Fan convectors mounted adjacent to walls shall be rigidly secured to the wall in accordance with the manufacturer's recommendations. Each fan convector shall incorporate a cleanable type air filter. Convector casings and coil fins are to be protected from damage, and left in a complete and clean state.

THERMAL INSULATION:

GENERAL: carry out thermal insulation work using one of the scheduled firms employing skilled craftsmen conversant with class of work.

Do not apply thermal insulation until installation has been fully tested and all joints proved sound.

Ensure materials are kept dry.

SEPARATION: insulate each unit separately. Do not enclose adjacent units together.

CLEARANCE: ensure adequate clearance between insulated pipes.

APPLICATION: apply insulants, facings, coatings and protection strictly in accordance with manufacturer's instructions.

© NSR 2008 - 2009

FINISH: neatly finish joints, corners, edges and overlaps and, where possible, arrange overlaps to fall on blind side. Ensure overlaps are neat and even and parallel to circumferential and longitudinal joints.

FLANGES AND VALVES: cut back to allow removal of bolts and nuts, finish with neat bevel or use end caps. Where boxes are used fit over insulation on adjacent piping.

LINERS: where load bearing insulation is required use segmental liners suitable for temperature. Fit insulant up to liner and carry facings across the pipe ring.

PREFORMED RIGID SECTIONS: use preformed sections formed in half pipe units or segments approximately 1000mm long to thickness required.

Factory sections	:	Use factory made sections wherever possible for fittings to suit required configuration and to same detail as insulation on straight runs
Preformed sections	:	Use preformed sections for insulating circular and curved surfaces of ductwork or of plant and equipment, to suit curvature
Shape sections	;	Use shape sections for insulating angle flanges on ductwork to suit locations and maintain condition of adjacent insulants across flange. Alternatively increase insulation thickness to incorporate flanges within insulation

PIPELINE SUPPORTS: insulation carried through between pipe and supports

: For load bearing insulation, carry through insulation and finish, and provide a 2mm thick sheet metal protective sleeve (for installation by pipe erector)

: For non-load bearing insulation on hot piping, close butt to a section of load bearing finished material 100mm long and provide a 2mm thick sheet metal protective sleeve (for installation by pipe erector)

: For non-load bearing insulation on cold piping, close butt to preformed hardwood segments, rigid polyurethane or high density phenolic foam inserts 100mm long. Carry over the finish 40mm and tape joint to maintain vapour barrier. Provide a 2mm thick sheet metal protective sleeve (for installation by pipe erector)

Insulation not carried through between pipe and support

: Provide end caps to match applied finish

PIPE SLEEVES: carry finished insulation through pipe sleeve. Pack annular space between insulation finish and sleeve with non-flammable and fire retardant material to form fire stop.

VALVE AND FLANGE BOXES: fabricate box in two sections, hinged together at one side and fastened with neat spring clip fasteners at the other. Make box to ensure that:-

: Insulation is of such thickness as permit easy fitting and removal of box

: Insulation fits tightly against the base without gaps

: Operation of valve remains unimpaired with box in place

VAPOUR BARRIERS – LIQUID: apply vapour seal solution evenly by brush in accordance with manufacturer's instructions; use solution which dries to a colour distinctive from insulating material.

VAPOUR BARRIERS – SHEET: apply membrane, sheet or laminated vapour barrier materials in continuous form. Otherwise return barrier to surface and effectively seal. Fit removable portions of insulation to access doors etc. as separate items with vapour barriers overlapping and sealed to main vapour barrier.

INTEGRITY OF VAPOUR BARRIERS: where a vapour barrier is indicated take particular care to ensure its integrity throughout.

Repair immediately any damage to vapour barriers and where such barriers have been applied off site, repair to specific manufacturer's instruction.

Where aluminium sheeting is used for protection submit proposals for securing sheeting without impairing the integrity of the vapour seal for approval.

PROTECTION: ensure that where protection is applied to insulation the joints fall blind side and that all joints are made to shed water and sealed with waterproof tape, adhesive or joint sealant where appropriate.

APPLICATION OF WEATHER PROOF PAINTS: do not apply weather proof paints in ambient temperatures below 5°C.

INSPECTION AND TESTING: arrange performance test of thermal conductivity on materials selected, carried out at manufacturer's works or at an approved laboratory and in accordance with appropriate British Standard, or as directed.

Apply additional insulation to meet specified values if tests demonstrate a lesser thickness than that dictated by thermal conductivity specified.

THICKNESS OF INSULATION: as specified.

BOILERS:

Install all equipment on purpose made bases or supports, capable of carrying load of equipment. Comply with the manufacturer's installation instructions.

Install all site constructed equipment to manufacturer's instruction. Ensure joints are pressure tested prior to installation of lagging and casings.

Replace all damaged equipment, lagging and casings.

UNIT HEATERS:

Each unit shall be installed such that all air venting can take place through the adjacent pipe connections, and via an extended air pipe with cock where this is not possible.

CONTROL COMPONENTS:

GENERAL: install pipeline control components to comply with BS ISO 9000 and manufacturer's instructions. Install ductline control components in accordance with DW 142 and manufacturer's instructions.

APPEARANCE: arrange, support and clip all control wiring, pneumatic tubes and capillaries to present a neat appearance, with other services and the building structure.

INSULATION: where control components are incorporated in insulated pipelines, ductlines or equipment, provide details for approval of method proposed to insulate component.

SUPPORTS: arrange supports for control components to ensure no strain is imposed on components.

ACCESS: arrange control components to ensure adequate access for operation and maintenance.

PRESSURE REDUCING AND CONTROL VALVES: install valves in horizontal pipelines and include moisture traps, strainers, pressure gauges and safety valves according to application and to the manufacturer's recommendations.

SELF OPERATING CONTROLS: install self operating controls in accordance with manufacturer's instructions and relevant standards. Generally install valves horizontally.

POWER OPERATED CONTROLS: install power operated controls in accordance with manufacturer's instructions and relevant standards.

ELECTRIC MOTOR ACTUATORS: securely mount actuators to rigid members, free from vibration or distortion. Select mounting positions to require minimum linkages, and to avoid angular drive to operating levers. Allow access for servicing and replacement.

SENSORS/ CONTROLLERS: install sensors/controllers in accordance with manufacturer's instructions, in accessible locations. Install wall mounted components, where indicated.

ANCILLARIES: install ancillaries in accordance with manufacturer's instructions.

ENCLOSURES: install enclosures where indicated, ensuring adequate space for access and maintenance.

DUCTWORK ACCESSORIES AND COMPONENTS:

WORKMANSHIP GENERAL: install ductwork in accordance with BS 5720, DW 142, DW 151 and DW 191 as appropriate.

ACCESS: where man access is provided, ensure that duct floor is of sufficient strength to comply with safety standards.

DRAINAGE OF DUCTWORK: arrange ductwork to drain any entrained moisture and ensure the lapping of joints prevents moisture leakage.

CONNECTION TO BUILDERS WORK: comply with DW 142 Part 7 Section 29.

SPACING SUPPORTS: for the spacing of supports comply with DW 142 Part 6, DW 151 Section 7 or DW 191 Section 7 as appropriate.

DUCTWORK VIBRATION ISOLATION: ensure that ductwork does not come in direct contact with building fabric except in cases of fire dampers, silencers and builder's frames. Isolate all supporting members from ductwork, secure to support by means of adhesive:-

    Lining of 6mm thick rubber

FLEXIBLE DUCTWORK: ensure that flexible ductwork does not become kinked or flattened. Support flexible ductwork using wire looped tie supports to prevent sagging.

PROTECTION AND CLEANING: cover open end of ducts during installation period to prevent ingress of dirt and rubbish by use of galvanised sheet metal cap ends or polyethylene sheeting. Clean each range of ductwork during progress of installation.

WATERPROOFING: Fit ductwork with trimming angle and weather cravat, skirt, flashing plate and cowl where ductwork passes through or terminates in roof, to ensure a weatherproof seal to building structure, as indicated.

DUCTWORK SLEEVES: enclose all ducts passing through building elements, (walls, floors, partitions, etc.) within purpose made sleeves. Cut sleeves of the same material as the duct and pack with mineral fibre or similar non-flammable and fire resistant material to form a fire/ smoke stop of adequate rating and to prevent air movement and noise transmission between duct and sleeve.

Provide flanges on either side of wall where ductwork is exposed in rooms.

TEST HOLES: provide test holes in ductwork system to allow complete testing and balancing of system in accordance with CIBSE Commissioning Code Series A. Drill test holes on site in accordance with DW 142 Part 7 Section 21.4 or DW 151 Section 10.5 as appropriate. Provide test holes:-

1. One each side of all equipment in system

2. At least 1.5 duct diameters upstream of all dampers

3. Provide, at each location, the number of test holes shown below:-

Circular ducts

Duct diameters	No. of test holes
Up to 150mm	1
151 to 450mm	2
Over 450mm	4

Rectangular ducts

Longest side	No. of test holes
Up to 200mm	1
201 to 400mm	2
401 to 600mm	3
601 to 800mm	4
801 to 1000mm	5
1001 to 1200mm	6
1201 to 1500mm	7
Over 1500mm	one per 250mm

HOLES FOR CONTROL EQUIPMENT: provide holes in ductwork to accommodate thermostats, humidistats and other control sensors as indicated. Where holes are provided on insulated ductwork, extend to finish flush with insulation.

INSTALLATION OF CONTROL EQUIPMENT: fit sensors, damper motors and other control equipment as indicated.

DUCTWORK AIR LEAKAGE TESTING: test low and medium pressure ductwork in accordance with DW 142 and DW 143.

DUCTWORK ANCILLARIES

CONSTRUCTION AND FINISHES: ensure that materials of accessories are compatible with ductwork and that finishes of accessories comply with any special requirements for ductwork.

ACCESS COVER RESTRAINT: install restraining straps to avoid loss of covers.

ACCESS DOOR HANDLES: provide handles on all access doors.

ACCESS OPENING SAFETY: provide safety bars where indicated at the top of risers.

ACCESS DOOR INSULATION: insulate access doors as ductwork.

DUCTWORK SUPPORTS: support ductwork in accordance with DW 142 Part 6 Section 19, DW 151 Section 7 or DW 191 Section 7 as appropriate.

VAPOUR BARRIER: Where a vapour seal is specified, to ensure continuity over ductwork support use:-

> Method 1 of DW 142 Part 6 Section 19.6.1

ACCESSORY SUPPORT: in accordance with DW 142 Part 6 Section 19.5 for supporting ductwork ancillaries:-

> Provide supports as specified

Provide additional supports adjacent to dampers, diffusers and other items of equipment to prevent distortion.

EXTERNAL DUCTWORK SUPPORT: support ductwork external to buildings as indicated.

APPEARANCE OF DUCTWORK SUPPORTS: cut off protruding ends of hanger rods and bolts close to nuts. Ensure that supports and drop rods are clear of ducts and not enclosed in thermal insulation, as DW 142 Section 19.6.

FIRE PRECAUTIONS: install fire dampers as indicated.

FIRE DAMPER ACCESS: ensure adequate access to fire damper mechanism through access doors, false ceilings etc. Demonstrate that fire damper blades close completely and that fire links can be replaced. Where more than one fire damper is installed in a frame ensure access is provided to all fire dampers.

POSITIONING: position components as indicated and in accordance with manufacturer's instructions.

INSTRUMENT CONNECTIONS: provide instrument connections where indicated.

TESTING: test ductwork components with the ductwork.

SECTION H

MEASUREMENT AND PRICING PREAMBLES

FORMAT OF DESCRIPTIONS

In addition to common abbreviations the following have been adopted:-

m	-	Metre
mm	-	Millimetre
m^2	-	Square metre
mm^2	-	Square millimetre
nr	-	number
N	-	Newton
t	-	Tonne
l	-	Litre
C	-	degrees Celsius
BS	-	British Standard
CP	-	Code of Practice

Other metric symbols are given in accordance with BS 6430.

Descriptions are usually given in the plural but secondary phrases which qualify or describe work in addition to the main part of a description may be in the singular.

Every description is to be read as if the phrase "and the like" were incorporated in it.

In order to avoid future amendments, these Preambles may contain references to items which may not necessarily be represented by rates in the Schedule.

The preambles contained in this Section are for guidance only to indicate the basis on which the level of pricing has been made.

All standards referred to within these documents shall be held to be the latest edition published at the date of tender.

A reference to any Act of Parliament or to any Order, Regulation, Statutory Instrument, Building Standard, Code of Practice or the like shall include a reference to any amendment or re-enactment of the same.

DEFINITIONS

"Approved", "Directed", "Selected" and similar expressions shall relate to the decision of the Contract Administrator.

"Works by Others", "Builders work", "Prepared by Others" and similar expressions relate to work done before or to be done during the current contract by Contractors, Sub-Contractors and public bodies directly employed by the Employer.

"Existing": in existence prior to the current contract.

PRICES ALSO TO INCLUDE

"Prices also to include", items which are to be included in the Contractor's Percentage adjustment fall into two categories:-

a) those which appear under the headings "Prices also to include" for which the Contractor is to allow in his Percentage adjustment

b) those which are deemed to be included in descriptions by the method of measurement and are not therefore specifically mentioned but which the Contractor is also obliged to include in his Percentage adjustment.

"Protect the work". Temporarily casing up, covering, protecting and the like to ensure that the work is left clean and perfect at the completion of the works.

"Cut" to include drilling or executing the labour described in any other way.

"Fix Only" shall mean receiving goods and materials, transporting to site, unloading, hoisting, unpacking, assembling, storing, site distribution and positioning, returning packing materials as applicable to the supplier carriage paid and obtaining credits thereon in addition to fixing to any backgrounds and connection.

"Supply Only" shall mean the nett invoice costs of goods and materials, including carriage less the defrayment of all trade and preferential discounts and credits for packing materials returned to the supplier.

"Connect Only" shall include dismantling, connection and re-assembling equipment supplied and fixed by others.

All fixings are to include their penetration of any intervening soft materials.

SCAFFOLDING

NOTE: the scaffolding rates in this section allow for basic forms of scaffolding. Where more complex forms of scaffolding are required, sub-contract rates should be used.

Scaffolding will only be paid for in accordance with this section when the height of the working platform exceeds 1.50m above ground level. Independent and putlog scaffolds are to be measured at the perimeter of the structure to be scaffolded x the height of the scaffold to comply with the current appropriate regulations.

The rates in this section allow in each item for the cost of scaffolding from ground level to the working platform including boards and double guard rails at one level only.

MECHANICAL SERVICES

DEFINITIONS

"Fixed" includes fixing with nails, screws, plugs, shot firing pins, bolts, ragbolts, expansion bolts, self drilling anchors, nuts, spring nuts, self locating nuts etc.

"Pipes" includes pipes and tubes.

"Bends" includes elbows, easy bends and slow bends both manufactured and made on site of varying degrees and radii.

"Reducers" includes excentric and concentric patterns.

"Brackets" includes standard supports (e.g. clips, saddles, rings, holderbats, hangers) together with component parts.

"Connector" includes any thimble, ferrule, one caulking bush, cap and lining or other fittings or adaptor whether equal or reducing or a combination of any of the above necessary to achieve the connection described.

PRICES TO INCLUDE

Generally

Pipework: pipes are deemed to include joints in their running length and are measured including brackets.

Pipework fittings and ancillaries: where these are to be enumerated separately, cutting and joining pipes to the fittings and ancillaries including reducers are deemed to be included.

Pipe Line Equipment: rates are deemed to include for all necessary bracketry and fixings, and everything necessary for jointing. Anti vibration mountings are to be measured separately.

Thermal Insulation: pipework fittings, ancillaries and supports are included in the straight run up to the specified pipe size, over that size fittings and ancillaries are to be measured separately.

NOTE

The pipework rates in the Schedule are based on fixing to timber or masonry using munsen rings or similar. The rates do not include for drop rods, unistrut, gallows brackets or angle iron supports.

Hot and Cold Water Services

Cylinders:	Vertical, pre-insulated, mounted on builders work base,
Circulation pumps:	Fabricated angle iron bracket, 2 coats red oxide, fixing to wall. Pump mounted on bracket, complete with union connections.
Storage tanks:	Positioning of tank on builders work base. Fixing ball-valve, fixing cover, applying thermal insulation.
Water heaters:	All connections, both electrical and mechanical.
Pipework:	Fixing to building structure using self-drill anchor, studding and ring clip. Bracket spacing as defined in the current edition of the CIBSE Guide

Fittings:

Up to and including 28mm copper	Solder ring capillary type
35mm to 54mm	Copper compression
67mm and over	Flanged fittings

Compression joints:

Connection to items of new or repaired equipment, valve, etc	Fitting new compression olive, cutting back, re-jointing, providing short length of pipe as necessary

Flanged joint:

Connection to item of new or repaired equipment, valve, etc	Providing new joint ring, re-using existing flange nuts, bolts and washers
	Item assumes that new equipment may be installed in the same space as that previously occupied by existing
Repairs to leakage pipe or fitting:	Re-tightening, re-soldering, wiping, re-making joint. No modification of pipework
Remove damaged pipework and fittings:	Cutting out and replacing 1m of damaged pipework, providing 2 nr new joints/ fittings
	or
	Disconnecting, removing and replacing 1 m of damaged pipework between 2 nr existing joints/ fittings, replacing with new

Copper pipework:

Up to and including 54mm	Compression type fittings, (or capillary up to 28mm) Munsen rings fixed to skirting @ 1 per m.

Hot and Cold Water Services (Cont'd)

67mm and over	Flanged joints. Includes new joint ring, nuts, bolts and washers with new flanges. Fixed at low level. Brackets @ 1 per m,
Galvanised pipework:	BS 1387 Medium weight galvanised mild steel fixed at low level, brackets, @ 1 per m
Valves and Ancillaries:	Various manufacturers
Up to and including 50mm	BSPT screwed connections
Up to and including 54mm	Copper compression connections
Ballofix valves	Brass finish
Air bottle assembly	50mm diameter, 200mm long, with 15mm drain connection including 2.5m of pipework, 2 nr fittings and 15mm lockshield valve, fixing with wall clip at spacing as defined in the current edition of the CIBSE Guide
UPVC pipework:	One coupling per metre
Automatic air vent	Includes 15mm drain connection including 2.5m of pipework and 2 nr fittings fixed with wall clip at spacing as defined in the current edition of the CIBSE Guide
Connections to equipment:	
Showers & mixing valves:	Based on "Mira" and other ranges
Laboratory taps, water heaters, etc	Copper compression type fittings
	Laboratory taps based on various manufacturers
Thermal Insulation:	
Internal Specification: No metal cladding	Mineral wool, foil face, rigid section. Cold water vapour sealed. Valves, etc included in straight run
Making good damaged thermal insulation	Removing 1m of damaged thermal insulation and replacing with new of similar type. Vapour seal to be maintained for cold water insulation

Space Heating and Services

Boilers:	Positioned on level base prepared by others.
Balanced flue	Fixing through builders work opening and seal
Conventional flue	Including 135° tee with drain, flat flashing and rain cap, passing through roof builders work opening
	Wall brackets spaced in accordance with manufacturer's recommendations
Convectors:	
Concealed type. Natural or fan-assisted	Postioning on level based prepared by others. Fix ducts 280mm long for supply and return air through builders work opening, affix supply and return air grilles
Fan convectors	FCU inclusive of plinth, filters, L.T thermostat, speed control, s/w switch, casing, finishes.
Flanged connections:	
New valves, compensators, etc	Providing new joint ring. Re-use existing flange, nuts, bolts and washers. Item assumes that new equipment may be installed in same space as that previously occupied by existing
Unit Heaters:	
Wall mounted type	Fabricated angle iron cantilever type bracket, 2 coats red oxide, fixing to building structure using self-drill anchors/ bolts and washers, mounting of unit heater on bracket
Ceiling mounted type	Fixing to soffit using self-drill anchors, 4 nr studding drop rods 1m long fixed to unit heater lugs with nuts and washers
Radiant panels	
Feed and expansion tanks	Fabricated angle iron cantilever type bracket, two coats red oxide, fixing to building structure using self-drill anchors/ bolts and washers, mounting of tank on bracket (with board beneath polypropylene tank)
	Fixing ball valve, fixing cover, and applying thermal insulation

Space Heating and Services (Cont'd)

General:

Fixing/ refixing:

Steel/ cast iron radiators, Natural/ fan assisted convectors. Perimeter heating finned-tube element	Removal by closing of radiators/ convector/ heating Valves without draining of the system
Rubber compensators and expansion bellows	
Up to and including 50mm	BSPT screwed connections
65mm and over	Flanged connections. Includes mating flange joint ring, nuts, bolts and washers
Pipework:	Fixing to building structure using self-drill anchor, studding and ring clip. Bracket spacing as defined in the current edition of the CIBSE Guide
Repair leaking pipe or fitting:	Re-making or re-tightening joint. No modification of pipework
Remove damaged pipework and fittings:	Cutting out and replacing of 1m of damaged pipework, providing 2 nr new joints/ fittings
	or
	Disconnecting, removing and replacing of 1m of damaged pipework between 2 nr existing joints/ fittings, replace with new
Copper Pipework:	
Up to and including 54mm	Compression type fittings, (or capillary up to 28mm) Munsen rings fixed to skirting @ 1 per m.
67mm and over	Flanged joints. Include new joint ring, nuts, bolts and washers with new flanges: Brackets at low level
Steel Pipework:	BS 1387 medium weight black mild steel
Up to and including 50mm	1 bracket per m.
65mm and over	1 bracket per m

Space Heating and Services (Cont'd)

Ancillaries:

Isolating valves	)
Double regulating valves	)
Metering stations	)
Orifice valves	) Various manufacturers
Commissioning sets	)
Plug valves	)
Check valves	)
Strainers	)

Up to and including 50mm   BSPT screwed connection

65mm and over   Flanged connections. Includes mating flange, joint ring, nuts, bolts and washers

Air bottle assembly   50mm Diameter, 200mm long, with 15mm drain connections including 2.5m of pipework, 2 nr fittings and 15mm lockshield valve, fixed with wall clips at spacings as defined in the current edition of the CIBSE Guide

Automatic air vent   Includes 15mm drain connection including 2.5m of pipework and 2 nr fittings, fixed with wall clips at spacings as defined in the current edition of the CIBSE Guide

Fixing/ refixing existing:

Steel/ cast iron radiator	)
	) 2.5m of copper or mild steel. Pipework with 4 nr
Natural/ fan assisted convector	) fittings and 2 nr brackets. Cut back, re-joint. All
	) with isolating valves
Perimeter heating finned-tube element	)

Remove/ replace radiator/ convector valve: Replacing valve, re-setting valve position, draining and filling measured elsewhere

Thermal Insulation:

Internal Specification. No rigid metal cladding fittings   Mineral wool, foil face, rigid section. Valves and fittings up to and including 50mm are included in straight run

Making good damaged thermal insulation: Removing 1m of damaged thermal insulation and replacing with new of similar type and finish. Mineral wool, foil face, rigid section

Fire Protection Services

Dry Riser:

Landing valves	Include for counter-flanges, joint rings, nuts, bolts and washers
Inlet breeching valves	

Ancillaries:

Isolating valves	)
Globe valves	) Various manufacturers
Non-return valves	)
Up to and including 50mm	BSPT screwed connections
65mm and over	Flanged connections. Includes mating flange, joint ring, nuts, bolts and washers
Repair leaking pipe or fitting:	Re-making or re-tightening joint. No modification of pipework
Remove damaged pipework and fittings:	Cutting out and replacing 1m of damaged pipework, provide 2 nr new joints/ fittings
	or
	Disconnecting, removing and replacing 1m of damaged pipework between 2 nr existing joints/ fittings, replace with new
Damaged hose reel final connections:	Local draining of system, replacing final connection pipework and 2 nr fittings, including setting, 1 metre long, re-fill system
Copper pipework:	As hot and cold water services
Galvanised pipework:	As hot and cold water services

Thermal Insulation:

Internal Specification. No metal cladding. Valve and flange boxes priced as extras	Mineral wool, foil face, rigid section. Valves and fittings up to and including 50mm included in straight run

Making good damaged thermal insulation:

Internal Specification. No metal cladding	As hot and cold water services

Ventilation Ductwork

Ductwork: Fixing to building structure using self drill anchors, drop rods and brackets. Bracket spacing as DW 142 or DW 151 or DW 191 as appropriate

Flanges, sealants, gaskets and tapes included in the rate

Vibration isolation, test holes, holes for control equipment included in the rate

Ductwork fittings: Fittings are measured as extra over items to the ductwork rate

Gas services

Connections to equipment:

Gas heater, laboratory fittings, etc   Black mild steel tube with malleable iron screwed fittings. Union connection to appliance. Testing for leaks

Remove damaged pipework and fittings: Cutting out and replacing 1m of damaged pipework, providing 2 nr new joints/ fittings

or

Disconnecting, removing and replacing 1m of damaged pipework between 2 nr existing joints/ fittings, replace with new

Testing for leaks

Damaged laboratory/ bench final connections: Replacing final connection pipework and 2 nr fittings, including setting, 1m long

Testing for leaks

Plant room services

Remove pump/ fix:

Up to and including 50mm	Isolating valves line size, reducing to pump connection with union. Disconnecting/ re-connecting at union. Disconnecting/ re-connecting pump drain for floor mounted pump. Venting pump casing. Checking pump direction for rotation.
65mm and over	Isolating valves line size reducing to pump connection with flange. Disconnecting/ re-connecting at flange. Replacing joint ring, re-using flange nuts, bolts and washers. Disconnecting/ re-connecting pump drain. Venting pump casing. Checking pump direction of rotation.

Remove control valve/ refix:

Isolating/ double regulating valve/ metering station line size reducing to valve connections

Up to and including 50mm	Disconnecting/ re-connecting with unions at valve ports
65mm and over	Disconnecting/ re-connecting with flanges at valve ports
	Replacing joint ring, re-using flange nuts, bolts and washers

Burners

Fire valve accessories	Based on " National Vulcan" range

MECHANICAL

PART 2

Save money and complete your financial management jigsaw

ABOUT US

For over 12 years NSR Management have been promoting and developing the National Schedule of Rates on behalf of the Co-authors, the Society of Construction and Quantity Surveyors and the Construction Confederation.

The National Schedule of Rates have become synonymous with the effective management of Measured Term Contracts, thus enabling organisations to be more efficient and save money.

THE NATIONAL SCHEDULE OF RATES

The National Schedule of Rates are in use throughout the whole of the UK, helping hundreds of organisations in both the public and private sector manage and maintain their diverse property needs.

They have also become an invaluable tool for benchmarking and ensuring that planned maintenance agreements are operated as cost effectively as possible.

NSR Management offer a whole range of National Schedule of Rates in book, data and internet subscription format.

The Schedules comprise, in total, approximately 16,000 items of work, covering Housing Maintenance, Building, Mechanical, Electrical, Access Audit, Roadworks and Painting and Decorating.

FINANCIAL BENEFITS FOR YOU AND YOUR ORGANISATION

- **Keeping Costs Down** — More competitive tenders are obtainable from contractors as they become more familiar with the schedules and the level of pricing within them.
- **Information Up Front** — By instructing contractors to use our schedules you will have the benefit of 'up front' cost estimation.
- **Piece of Mind** — The schedules are in use throughout the whole of the UK. and are updated annually to ensure the information is a useful as possible.
- **Good Benchmarking** — The use of our schedules is extensive, hence they are an ideal benchmarking **tool**.

AND FINALLY

NSR Management offer training in the use of the National Schedule of Rates and our 'One Stop Shop' consultancy service is available to assist with contract advice, benchmarking, QS services, condition surveys, contract management and DDA surveys.

Our after sales service is second to none, well that's what our customers tell us, so technical support is on hand for all of our schedules, or any of their associated computer applications.

After all *'Our Business is Your Property'*.

Please contact us for more information

NSR Management Limited
Pembroke Court
22-28 Cambridge Street
Aylesbury, Bucks HP20 1RS

tel: 01296 339966
fax: 01296 338514

nsr@nsrmanagement.co.uk

www.nsrm.co.uk

GENERAL INDEX (PART 2)

	Page
Index	2/1
Summary of Amendments	2/5
Compilers Notes	2/6
Schedule of Descriptions and Prices	2/17
Basic Prices: Labour, Plant and Materials	BP

Our Business is your Property

On Site Training in the use of National Schedules

Our standard full day training course 9.30am – 4.30pm comprises the following sessions:

- Benefits of Measured Term Contract
- Estimate Percentage 'A'
- JCT Measured Term Contract
- Communication
- Dispute Resolution
- Some Practical Problems
- Estimating the Value of Works Orders

We usually suggest that numbers should be restricted to no more than 8 and the total cost for the training is £1250 [+VAT] plus speaker's expenses which should not exceed £200. If more that 8 attend the cost for each additional delegate is £200

You could always invite others from out with your company to defray costs if you have insufficient in house staff who wish to participate.

Shorter sessions can be organised to suit your particular needs and the cost would be adjusted accordingly.

NSR Management Ltd
Pembroke Court, 22-28 Cambridge Street, Aylesbury, HP20 1RS   Telephone: 01296 339966   Facsimile: 01296 338514
e-mail: nsrm@nsrmanagement.co.uk   www.nsrm.co.uk   Est. 1995   Registered in England 3574827   VAT No. 640 0550 83

INDEX

Listed below are the main Section Titles and Library Groups which have been used in the formulation of the Schedule

NSR MECHANICAL CODING

NSR Coding A

44	Contractors General Cost Items

NSR Coding C

83	Piped supply systems hot and cold water

NSR Coding P

31	Builders Work: Chases and Holes

NSR Coding R

10	Plumbing: Soil and Waste

NSR Coding S

15	**Hot and Cold Water Services : Plant and Equipment**
1500	Cylinders
1518	Circulating Pumps
1527	Showers and Mixing Valves
1548	Storage Tanks
1575	Laboratory Fittings
17	**Hot and Cold Water Services : Pipework and Ancillaries**
1700	Pipework
1730	Valves and Ancillaries
1784	General
25	**Space Heating Services : Plant and Equipment**
2500	Feed and Expansion Tanks

27		**Space Heating Services : Pipework and ancillaries**
	2700	Pipework
	2730	Valves and Ancillaries
60		**Thermal Insulation**
	6000	Pipework
62		**Fire Protection Services**
	6200	Hose Reels
	6228	Dry Risers
	6260	Pipework
	6278	Valves and Ancillaries
76		**Laboratory Fittings**
	7620	Pipework
	7652	Valves and Ancillaries
85		**Water Treatment**
	8500	Water Softeners
	8502	Water Scale Inhibitors
90		**Sump and Submersible Pumps**
	9006	Sump Pumps
	9010	Submersible Pumps

NSR Coding T

15		**Hot and Cold Water Services**
	1500	Water Heaters
25		**Space Heating Services : Plant and Equipment**
	2510	Air Curtains
	2518	Boilers
	2520	Circulating Pumps
	2522	Radiators

2538	Convectors
2574	Perimeter Heating
2594	Unit Heaters
2597	Radiant Panels
26	**Space Heating Services : Programmers and timers**
27	**Space Heating Services : Pipework and Ancillaries**
2700	General
2748	Radiator Ancillaries
2780	Convector Ancillaries
2788	General
2789	Draining Down of Non-Domestic Heating System
55	**Refrigeration**
5500	Pipework
5524	Refrigerant Gas
5536	Unit Parts
5554	Component Parts
76	**Gas Services**
7600	Gas Heaters
80	**Plant Room Services**
8000	Circulating Pumps
8020	Control Valves
8038	Gas Burners and Gas Trains
8050	Oil Burners
8062	Fire Valve Accessories
8078	Valves and Ancillaries

NSR Coding U

32	**Ventilation Ductwork**
3200	Straight Ductwork
3220	Ductwork Fittings
3260	Grilles and Diffusers
45	**Air Conditioning**
4500	Fan Coil Units
4530	Compressors
4566	Air Conditioning Units

SUMMARY OF AMENDMENTS

The alterations to the Schedule are as follows:

New items for condensate pumps have been added as section U4590, U4592 and U4594.

The method of measurement for pipework has been changed to identify fittings separately as per SMM. The range of sizes has also been increased.

COMPILER'S NOTES

The National Schedule for mechanical Services is principally designed for use in the Public Sector, but is now being widely used in the private sector on work with a maintenance or refurbishment bias.

NOTE: **The Electrical and Plumbing Services contained in the Building NSR are for works comprising small domestic repairs and modifications forming a minor part of the building works.**

The rates contained in the Mechanical and Electrical Schedules are for direct engineering services works of a more substantial nature.

USERS ARE REMINDED THAT THE NATIONAL SCHEDULE IS NOT A PRICING BOOK; FURTHER, THEY ARE ADVISED TO TAKE A SELECTION OF RATES IN THE NATIONAL SCHEDULE AND TEST THEM AGAINST THEIR OWN DATA ON A REGULAR BASIS.

Effective Date of Schedule

The Schedule is revised each year to include latest promulgated wage rates and prices for materials and plant

User's attention is drawn to base dates for labour and materials given in the Compiler's notes and to the listed basic prices.

Users are advised that the rates in this Schedule are deemed to be effective as from 1st August 2008.

CALCULATION OF LABOUR RATES

Plumbers

The productive hours worked by plumbers have been taken as for the Building Operatives.

Hour worked in full year	1802	1802
Less inclement weather 2%	36	
Productive hours	1766	
Non-productive overtime		75
Public holidays (8 x 7.8 hours)		62.5
Hours for which payment is made		1940

	Advanced Plumber £	Trained Plumber £
Basic rate per hour	12.12	10.39
Additional payments for responsibility	0.44	N/A
Sub-total	12.56	10.39
Add bonus 30%	3.768	3.117
Total Cost per hour	£16.328	£13.507

Plumbers

Build-up of all-in labour rates per annum

			Advanced Plumber £	Trained Plumber £
Earnings	1940	x £16.328	31,678.32	-
Hours		x £13.507	-	26,203.58
Sub-total (A)			31,678.32	26,203.58
National Insurance: secondary contribution (Employer) (rates as from April 2005)				
0% on first £97.01 per week			-	-
12.8% on remainder	12.8% on £26,638.80		3,408.37	
	12.8% on £21,159.06			2,708.36
Industry pension 6.5% of (A)			2,058.96	1,703.23
Annual holiday credit and sickness benefit stamp				
52 weeks	x £35.10		1,825.20	
	x £31.21			1,622.92
Sub-total			38,969.35	32,238.09
Allowance for severance pay		2%)	779.39	644.76
Employer's liability		2%)	779.39	644.76
Trade supervision		6%)	2,338.16	1,934.29
Total cost per annum		£	42,866.29	£ 35,461.90
Rate per hour (divide by productive hours 1766)			£24.27	£20.08

Electricians

The hours worked by Electricians are based on a standard working week without any overtime, as follows:

Hour worked in full year	1732.50	17.32.50
Less inclement weather 2%	34.65	
Productive hours	1697.85	
Public holidays (8 x 7.5 hours)		60.00
Hours for which payment is made		1792.50

	Electrician £	Electrician's Labourer £
Basic rate per hour	13.08	9.67
Add bonus 30%	3.924	2.90
Total Cost per hour	£ 17.004	£12.57

Electricians

Build-up of all-in labour rates per annum

			Electrician £	Electrician's Labourer £
Earnings	1792.5	x £ 17.004	30,479.67	-
Hours		x £ 12.57	-	22,533.52
National Insurance: secondary contribution (Employer) (rates as from April 2005)				
0% on first £97.01 per week		-	-	-
12.8% on remainder		12.8% on £25,435.15	3,255.70	
		12.8% on £17,485.88		2,238.59
Industry Pension 6.5%			1,981.18	1,464.68
Combined benefit scheme 52 weeks				-
wks 3,4,5&6 £75, wks 7-28 £150		Elect	3,450.00	
wks 3,4,5&6 £70, wks 7-28 £140		Lab	-	3,220.00
Total			39,166.55	29,456.79
Allowance for severance pay		2%	783.33	589.14
Employer's liability		2%	783.33	589.14
Trade supervision		6%	2,349.99	1,767.41
Total cost per annum			£ 43,083.20	£ 32,402.47
Rate per hour (Divide by productive hours 1698)			£ 25.38	£ 19.08

MATERIAL COSTS

The costs of materials are based on standard published price lists current at March 2008. It is emphasised that the prices thus obtained are indicative only and the user should satisfy himself as to their validity.

No account has been taken of discounts available under annual purchasing agreements.

The prices are based on products marketed by various well-known medium priced manufacturers.

QUANTITIES, RATES, EXTENSIONS AND TOTALS

All quantities, rates, extensions and totals are to be kept to two decimal places.

DESCRIPTIONS

Definitions of the methods of measurement used are contained within Part 1, to complement the descriptions within the schedule itself.

COMPUTER-BASED LIBRARY

The schedule is fully computer based. This means that items of work are produced by using the computer to generate descriptions from a custom-compiled Library placed in its memory. This Library is divided into coded sections.

The benefits from using a Library are:

> the codes are identifiable in the Schedule : coding is permanent, compact and does not change with amendments or updates to the Schedule.

> several possibilities for amending the Schedule are available in order to add data, change existing data, etc.

COMPILER'S NOTES – APPENDIX A

COMPUTER-BASED LIBRARY

The Library for the Mechanical Schedule is divided into SMM7 (A, C, P, R, S, T and U) main Work Groups, as follows:

A : Contractors General Cost Items

C: Demolition/Alteration/Renovation

P: Building Fabric Sundries

R : Plumbing

S : Piped Supply Systems

T : Mechanical Heating/Cooling/Refrigeration Systems

U : Ventilation/Air Conditioning Systems

Each of these is divided into a number of Library Groups, for example:

A44 : Contractors General Cost Items

C83 : Piped supply systems

P31 : Holes/chases/covers/supports for plumbing or mechanical services

R10 : Soil and Waste

S15 : Hot and Cold Water Services

T25 : Space Heating Services : Plant and Equipment

U32 : Ventilation Ductwork.

Each of these is then divided into Combined Headings, for example:

S1527 : Showers and Mixing Valves

S1560 : Storage Tanks

These are then divided into Work Heads (item descriptions) as follows:

010	:	Remove damaged shower spray head; fix only new head
011	:	Remove, supply and fix shower spray head
012	:	Supply and fix new shower head
014	:	Remove damaged rigid shower assembly, fix only new assembly

©NSR 2008 – 2009

An example of the way in which these codes and texts appear in the Schedule is as follows:

S : MECHANICAL SERVICES WORKS OF ALTERATION/SMALL WORKS/REPAIRS.

S15 : Hot and Cold Water Services : Plant and Equipment (Library Group S15)

S1527 : Showers and Mixing Valves (Combined heading S1527)

010 : Remove damaged shower spray head, fix only new head (Work Head 010)

011 : Remove, supply and fix new shower spray head (Work Head 011)

012 : Supply and fix new shower spray head (Work Head 012)

014 : Remove damaged rigid shower assembly, fix only new assembly (Work Head 014)

S1575 : Laboratory Fittings

900 : Bench Mounted Pattern

902 : Wall Mounted Pattern

904 : Remote Control Mounted Pattern

The Library contains a number of headings and items that for compactness it has been decided not to call up for use in the Schedule. These provide scope for future amendments to the Schedule.

Several possibilities for future amendment to the schedule are available:

- additional coding can produce an item not currently printed in the Schedule by a combination of Combined Headings (CH) and Work Heads (WH) already in the Library. Such an item could be said to be in the form

    existing Library Group/existing CH/<u>new</u> WH

    existing Library Group/<u>new</u> CH/existing WH

    existing Library Group/<u>new</u> CH/<u>new</u> WH

    a <u>new</u> Library Group could use any of the three foregoing combinations :

in every case a unique code will be produced so that the required new item goes into its proper place in the Schedule and retains its number permanently.

Users introducing their own descriptions into the Schedule for their own use, are advised not to use the omitted numbers in the published Schedule since these "missing" numbers could lead to an obvious conflict but can easily be avoided if users introduce their own references in the form of suffix letters.

e.g. S1527/010A to define a new item (unique to User) which requires to be entered into the Schedule by the User immediately following the standard published item coded S1527/010.

N.B.

Notes:

1. Costs are specifically attached to an identifiable source and can therefore be uplifted in line with the increase in price of that item when notified by the supplier.

2. Materials often comprise of multiple items.

3. Cables are priced from rolls of either 50 or 100 metre rolls and no allowance has been made for cutting or for cut lengths.

4. Many components are available in specified lengths and no allowance has been made for the cutting to the schedules measured unit of measure i.e. metre. (e.g. pipe is available in 3 metre lengths).

5. Where materials of a specified size are not available, alternative materials are substituted and the alteration to the schedule has been carried out.

6. Pricing policies of the various manufacturers used, especially cable manufacturers, distort the real cost of the materials by the various discounts against the different materials.

7. Where there are descriptions in the schedule which have materials that are no longer available or the materials have been superseded an indicative cost has been assigned as possible sources might still be available, but not listed.

8. Where items in the schedule which refer to BS numbers which have been superseded, the item is to be in accordance with the new number e.g. BS2871 should be read as BSEN1057

MECHANICAL

SCHEDULE OF DESCRIPTIONS AND PRICES

Global Schedule of Rates

The Global Schedule of Rates computer system can be used on Measured Term Contracts to which any of the NSR Schedule of Rates apply or any other Schedule of Rates. Addendum Schedules are easily added.

For a Works Order or Estimate, the Schedule is either searched on screen and description selected or the required code number is typed in, followed by full dimensions or quantity. Non Schedule codes, such as Star Rates, Net Rates, Dayworks, Invoices, etc., are easily added to a measure. When the completed Order is displayed, dimensions are squared, the resultant quantity is multiplied by the rate and a total Order value generated. Relevant percentages are automatically applied including the Contractor's percentage.

Many reports are available enabling tight control of the contract. These include;
Orders issued but not completed.
Orders issued, completed or billed within a given period.
Orders processed for a selected Building.
Orders completed either within or outside a defined period.
Interim Payments.
How often a particular Nsr code or section of codes has been used on a Contract.
Order printout using, instead of total rate, plant, labour or material rates.

Order details can also be exported to Microsoft Excel and Word or Emailed to another office.

The software will run on any of the Full versions of Windows.

For further information;

Email: john@barcellos.co.uk

Telephone: 0116 233 5559   Fax: 0116 233 5560

Address: Barcellos Limited, Sandbach House, 8 Salisbury Road, Leicester, LE1 7QR

Website: www.barcellos.co.uk

			Mat. £	Lab. £	Plant £	Total £
	A : MECHANICAL SERVICES - CONTRACTORS GENERAL COST ITEMS : WORKS OF ALTERATION/SMALL WORKS/REPAIR					
	A44 : CONTRACTORS GENERAL COST ITEMS					
A4410	**Erect and dismantle scaffolding**					
050	Independent tied scaffold	m2		9.50		9.50
060	Putlog scaffold	m2		7.49		7.49
070	Additional boarded platform (measured on plan)	m2		7.49		7.49
080	Brick guards	nr		1.15		1.15
085	Fans	m2		8.64		8.64
087	Roof edge double guardrail	m		6.34		6.34
090	Temporary roofs (Plastic tarpaulin covered)	m2		12.96		12.96
092	Temporary dust screens (Plastic tarpaulin)	m2		3.46		3.46
094	Rubbish chute	m		1.44		1.44
096	Scaffold Hoist	nr		4.32		4.32
A4415	**Erect and dismantle chimney scaffold ne 2.00 m girth**					
100	1.50 - 3.00 m From ground level	nr		43.20		43.20
105	3.00 - 4.00 m From ground level	nr		57.60		57.60
110	4.00 - 5.00 m From ground level	nr		72.00		72.00
115	5.00 - 6.00 m From ground level	nr		86.40		86.40
A4420	**Erect and dismantle chimney scaffold 2.00 - 3.00 girth**					
100	1.50 - 3.00 m From ground level	nr		64.80		64.80
105	3.00 - 4.00 m From ground level	nr		86.40		86.40
110	4.00 - 5.00 m From ground level	nr		108.00		108.00
115	5.00 - 6.00 m From ground level	nr		129.60		129.60
A4425	**Erect and dismantle tower scaffold to provide a working platform of 2.50 x 0.85 m**					
100	1.50 - 3.00 m From ground level	nr		21.60		21.60
105	3.00 - 4.00 m From ground level	nr		25.92		25.92
110	4.00 - 5.00 m From ground level	nr		29.38		29.38
115	5.00 - 6.00 m From ground level	nr		34.56		34.56

© NSR 01 Aug 2008 - 31 Jul 2009   A1

A : MECHANICAL SERVICES - CONTRACTORS GENERAL COST ITEMS : WORKS OF ALTERATION/SMALL WORKS/REPAIR			Mat. £	Lab. £	Plant £	Total £
A4430	Erect and dismantle tower scaffold to provide a working platform of 2.50 x 1.45 m					
100	1.50 - 3.00 m From ground level	nr		23.62		23.62
105	3.00 - 4.00 m From ground level	nr		28.51		28.51
110	4.00 - 5.00 m From ground level	nr		32.26		32.26
115	5.00 - 6.00 m From ground level	nr		38.02		38.02
A4435	Erect and dismantle lightweight aluminium access units					
200	Chimney scaffold unit to provide working platform to half of centre ridge stack	nr		8.64		8.64
205	Chimney scaffold unit to provide complete working platform around centre ridge stack	nr		17.28		17.28
230	Window access unit, 450 mm wide platform	nr		4.32		4.32
235	Window access unit, 600 mm wide platform	nr		5.18		5.18
240	Staircase access unit, 300 - 450 mm wide platform	nr		4.32		4.32
245	Staircase access unit, 600 - 675 mm wide platform	nr		5.18		5.18
A4440	Partially and temporarily dismantle and re-erect scaffolding for safety reasons					
050	Independent tied scaffold	m2		7.20		7.20
060	Putlog scaffold	m2		7.49		7.49
070	Additional boarded platform (measured on plan)	m2		7.20		7.20
080	Brick guards	nr		1.15		1.15
085	Fans	m2		8.64		8.64
090	Temporary roofs (Plastic tarpaulin covered)	m2		12.96		12.96
092	Temporary roofs (Plastic tarpaulin covered)	m2		3.46		3.46
A4445	Partially and temporarily dismantle and re-erect chimney scaffold ne 2.00 m girth for safety reasons					
100	1.50 - 3.00 m From ground level	nr		36.00		36.00
105	3.00 - 4.00 m From ground level	nr		48.96		48.96
110	4.00 - 5.00 m From ground level	nr		61.92		61.92
115	5.00 - 6.00 m From ground level	nr		73.44		73.44
A4450	Partially and temporarily dismantle and re-erect chimney scaffold 2.00 - 3.00 m girth for safety reasons					
100	1.50 - 3.00 m From ground level	nr		54.72		54.72
105	3.00 - 4.00 m From ground level	nr		73.44		73.44
110	4.00 - 5.00 m From ground level	nr		92.16		92.16

© NSR 01 Aug 2008 - 31 Jul 2009   A2

A : MECHANICAL SERVICES - CONTRACTORS GENERAL COST ITEMS : WORKS OF ALTERATION/SMALL WORKS/REPAIR			Mat. £	Lab. £	Plant £	Total £
A4450						
115	5.00 - 6.00 m From ground level	nr		109.44		109.44
A4455	Partially and temporarily dismantle and re-erect tower scaffold to provide a working platform of 2.50 x 0.85 m for safety reasons					
100	1.50 - 3.00 m From ground level	nr		18.14		18.14
105	3.00 - 4.00 m From ground level	nr		21.60		21.60
110	4.00 - 5.00 m From ground level	nr		24.77		24.77
115	5.00 - 6.00 m From ground level	nr		29.38		29.38
A4460	Partially and temporarily dismantle and re-erect tower scaffold to provide a working platform of 2.50 x 1.45 m for safety reasons					
100	1.50 - 3.00 m From ground level	nr		20.16		20.16
105	3.00 - 4.00 m From ground level	nr		24.19		24.19
110	4.00 - 5.00 m From ground level	nr		27.36		27.36
115	5.00 - 6.00 m From ground level	nr		32.26		32.26
A4462	Erect and dismantle temporary security fencing					
300	3500 x 2200 mm galvanised mesh panels, placed in precast concrete base feet, fixing with clamps	nr		2.80		2.80
A4465	Scaffolding - Hire charge (weekly rate)					
050	Independent tied scaffold	m2			3.58	3.58
060	Putlog scaffold	m2			2.71	2.71
070	Additional boarded platform (measured on plan)	m2			3.14	3.14
080	Brick guards	nr			0.62	0.62
085	Fans	m2			4.66	4.66
087	Roof edge double guardrail	m			2.53	2.53
090	Temporary roofs (Plastic tarpaulin covered)	m2			0.90	0.90
092	Temporary dust screens (Plastic tarpaulin)	m2			0.90	0.90
094	Rubbish chute	m			12.75	12.75
096	Scaffold Hoist	nr			120.00	120.00
A4470	Chimney scaffold ne 2.00 m girth - Hire charge (weekly rate)					
100	1.50 - 3.00 m From ground level	nr			73.36	73.36
105	3.00 - 4.00 m From ground level	nr			100.32	100.32
110	4.00 - 5.00 m From ground level	nr			127.29	127.29

© NSR 01 Aug 2008 - 31 Jul 2009   A3

A : MECHANICAL SERVICES - CONTRACTORS GENERAL COST ITEMS : WORKS OF ALTERATION/SMALL WORKS/REPAIR			Mat. £	Lab. £	Plant £	Total £
A4470						
115	5.00 - 6.00 m From ground level	nr			154.25	154.25
A4475	Chimney scaffold 2.00 - 3.00 m girth - Hire charge (weekly rate)					
100	1.50 - 3.00 m From ground level	nr			111.56	111.56
105	3.00 - 4.00 m From ground level	nr			138.52	138.52
110	4.00 - 5.00 m From ground level	nr			165.49	165.49
115	5.00 - 6.00 m From ground level	nr			182.01	182.01
A4480	Tower scaffold to provide a working platform of 2.50 x 0.85 m - Hire charge (Weekly rate)					
100	1.50 - 3.00 m From ground level	nr			113.50	113.50
105	3.00 - 4.00 m From ground level	nr			133.00	133.00
110	4.00 - 5.00 m From ground level	nr			152.50	152.50
115	5.00 - 6.00 m From ground level	nr			172.00	172.00
A4482	Tower scaffold to provide a working platform of 2.50 x 0.85 m - Hire charge (Daily rate)					
100	1.50 - 3.00 m From ground level	nr			56.75	56.75
105	3.00 - 4.00 m From ground level	nr			66.50	66.50
110	4.00 - 5.00 m From ground level	nr			76.25	76.25
115	5.00 - 6.00 m From ground level	nr			86.00	86.00
A4485	Tower scaffold to provide a working platform of 2.50 x 1.45 m - Hire charge (Weekly rate)					
100	1.50 - 3.00 m From ground level	nr			113.50	113.50
105	3.00 - 4.00 m From ground level	nr			133.00	133.00
110	4.00 - 5.00 m From ground level	nr			152.50	152.50
115	5.00 - 6.00 m From ground level	nr			172.00	172.00
A4487	Tower scaffold to provide a working platform of 2.50 x 1.45 m - Hire charge (Daily rate)					
100	1.50 - 3.00 m From ground level	nr			56.75	56.75
105	3.00 - 4.00 m From ground level	nr			66.50	66.50
110	4.00 - 5.00 m From ground level	nr			76.25	76.25
115	5.00 - 6.00 m From ground level	nr			86.00	86.00

© NSR 01 Aug 2008 - 31 Jul 2009   A4

	A : MECHANICAL SERVICES - CONTRACTORS GENERAL COST ITEMS : WORKS OF ALTERATION/SMALL WORKS/REPAIR		Mat. £	Lab. £	Plant £	Total £
A4490	**Lightweight aluminium access units - Hire charge (Weekly rate)**					
200	Chimney scaffold unit to provide working platform to half of centre ridge stack	nr			110.00	110.00
205	Chimney scaffold unit to provide complete working platform around centre ridge stack	nr			220.00	220.00
230	Window access unit, 450 mm wide platform	nr			90.00	90.00
235	Window access unit, 600 mm wide platform	nr			90.00	90.00
240	Staircase access unit, 300 - 450 mm wide platform	nr			90.00	90.00
245	Staircase access unit, 600 - 675 mm wide platform	nr			90.00	90.00
A4491	**Lightweight aluminium access units - Hire charge (Daily rate)**					
200	Chimney scaffold unit to provide working platform to half of centre ridge stack	nr			55.00	55.00
205	Chimney scaffold unit to provide complete working platform around centre ridge stack	nr			110.00	110.00
230	Window access unit, 450 mm wide platform	nr			45.00	45.00
235	Window access unit, 600 mm wide platform	nr			45.00	45.00
240	Staircase access unit, 300 - 450 mm wide platform	nr			45.00	45.00
245	Staircase access unit, 600 - 675 mm wide platform	nr			45.00	45.00
A4492	**Temporary security fencing - Hire charges (Weekly rate 1 to 3 weeks)**					
300	3500 x 2200 mm galvanised mesh panels, placed in precast concrete base feet, fixing with clamps	nr			7.02	7.02
A4494	**Temporary heating/lighting - Hire charge (weekly rate)**					
350	Festoon lighting	nr			32.00	32.00
360	Temporary heating (per room)	nr			19.00	19.00
A4495	**Mechanical Access equipment - Hire charge (weekly rate)**					
300	Compact scissor lift 7.8m	nr			360.00	360.00
A4496	**Mechanical Access equipment - Hire charge (Daily rate)**					
300	Compact scissor lift 7.8m	nr			216.00	216.00
A4497	**Scaffold Lighting/Security - Hire charge (Minimum hire & weekly rate)**					
300	Security floor lighting, per light - Set up and minimum charge	nr			50.00	50.00
310	Security floor lighting	nr			3.50	3.50

© NSR 01 Aug 2008 - 31 Jul 2009

A5

A : MECHANICAL SERVICES - CONTRACTORS GENERAL COST ITEMS : WORKS OF ALTERATION/SMALL WORKS/REPAIR			Mat. £	Lab. £	Plant £	Total £
A4497						
320	Scaffold alarm, Quad beam and passive 50m range - Set up and minimum charge	nr			234.00	234.00
330	Scaffold alarm, Quad beam and passive 50m range	nr			21.00	21.00

© NSR 01 Aug 2008 - 31 Jul 2009

			Mat. £	Lab. £	Plant £	Total £
	C : DEMOLITION/ALTERATION/RENOVATION					
	C83 : Piped supply systems, hot and cold water					
C8310	**Copper pipework**					
102	Stripping out ne 28mm pipework	m		6.69		6.69
112	Stripping out 35 to 54mm pipework	m		8.03		8.03
C8312	**Plastic pipework**					
100	Stripping out ne 25mm pipework	m		3.35		3.35
110	Stripping out 32 to 50mm pipework	m		4.02		4.02
120	Stripping out 75 to 100mm pipework	m		6.02		6.02
C8315	**Mild steel pipework**					
102	Stripping out ne 25mm pipework	m		8.03		8.03
112	Stripping out 32 to 50mm pipework	m		10.04		10.04
120	Stripping out 65 to 80mm pipework	m		13.39		13.39
C8330	**Lead pipework**					
112	Stripping out ne 50mm pipework	m		8.03		8.03
	C85 : Low temperature hot water/heating					
C8510	**Generally**					
102	Disconnecting and removing hot water cylinders	nr		40.16		40.16
112	Disconnecting and removing combination tanks	nr		50.20		50.20
122	Disconnecting and removing storage tanks, n.e. 60 gallon	nr		48.19		48.19
130	Disconnecting and removing storage tanks, 61 - 100 gallon	nr		53.55		53.55
140	Disconnecting and removing storage tanks, 101 - 500 gallon	nr		68.27		68.27
150	Disconnecting and removing circulating pumps	nr		13.39		13.39
160	Disconnecting and removing shower mixing valve	nr		25.10		25.10
170	Disconnecting and removing sump pumps	nr		10.04		10.04
180	Disconnecting and removing gas fired boiler 13 - 20kw	nr		110.44		110.44
190	Disconnecting and removing gas fired boiler 23 - 43kw	nr		130.52		130.52
202	Disconnecting and removing oil fired boiler 13 - 20kw	nr		130.52		130.52
212	Disconnecting and removing oil fired boiler 23 - 43kw	nr		140.56		140.56
222	Disconnecting and removing warm air curtain	nr		16.73		16.73

© NSR 01 Aug 2008 - 31 Jul 2009   C1

	C : DEMOLITION/ALTERATION/RENOVATION		Mat. £	Lab. £	Plant £	Total £
C8510						
232	Disconnecting and removing radiators	nr		6.69		6.69
	C87 : Air conditioning systems					
C8710	**Generally**					
100	Disconnecting and removing air conditioning system, inc reclaim refrigerant gas and store for reuse	nr		138.89	204.00	342.89
110	Remove compressor, inc dismantle unit	nr	2.60	138.89		141.49

© NSR 01 Aug 2008 - 31 Jul 2009   C2

			Mat. £	Lab. £	Plant £	Total £
	P : BUILDING FABRIC SUNDRIES					
	P31 : Holes/chases/covers/supports for plumbing or mechanical services					
P3150	**General builders work**					
300	Cutting or forming holes for pipes ne 55 mm nominal size in concrete 100 mm thick, making good	nr	0.37	3.88	4.56	8.81
302	Cutting or forming holes for pipes ne 55 mm nominal size in concrete 200 mm thick, making good	nr	0.37	4.74	5.96	11.07
304	Cutting or forming holes for pipes ne 55 mm nominal size in common brickwork 102 mm thick, making good	nr	0.56	3.02	3.56	7.14
306	Cutting or forming holes for pipes ne 55 mm nominal size in common brickwork 215 mm thick, making good	nr	0.56	3.23	3.99	7.78
308	Cutting or forming holes for pipes ne 55 mm nominal size in common brickwork 328 mm thick, making good	nr	0.56	4.31	6.01	10.88
310	Cutting or forming holes for pipes ne 55 mm nominal size in facing brickwork 102 mm thick, making good	nr	0.56	3.88	4.56	9.00
312	Cutting or forming holes for pipes ne 55 mm nominal size in facing brickwork 215 mm thick, making good	nr	0.56	4.31	5.21	10.08
314	Cutting or forming holes for pipes ne 55 mm nominal size in facing brickwork 328 mm thick, making good	nr	0.56	5.82	7.59	13.97
316	Cutting or forming holes for pipes ne 55 mm nominal size in blockwork 60 mm thick, making good	nr	0.37	1.29	1.45	3.11
318	Cutting or forming holes for pipes ne 55 mm nominal size in blockwork 75 mm thick, making good	nr	0.37	1.51	1.68	3.56
320	Cutting or forming holes for pipes ne 55 mm nominal size in blockwork 100 mm thick, making good	nr	0.37	1.72	1.96	4.05
322	Cutting or forming holes for pipes ne 55 mm nominal size in blockwork 150 mm thick, making good	nr	0.37	1.72	2.01	4.10
324	Cutting or forming holes for pipes ne 55 mm nominal size in blockwork 200 mm thick, making good	nr	0.37	1.72	2.06	4.15
326	Cutting or forming holes for pipes ne 55 mm nominal size in blockwork 215 mm thick, making good	nr	0.37	1.94	2.28	4.59
350	Cutting or forming holes for pipes 55 to 110 mm nominal size in concrete 100 mm thick, making good	nr	0.75	5.39	6.14	12.28
352	Cutting or forming holes for pipes 55 to 110 mm nominal size in concrete 200 mm thick, making good	nr	1.12	6.90	8.22	16.24
354	Cutting or forming holes for pipes 55 to 110 mm nominal size in common brickwork 102 mm thick, making good	nr	0.75	4.31	4.91	9.97
356	Cutting or forming holes for pipes 55 to 110 mm nominal size in common brickwork 215 mm thick, making good	nr	1.12	5.39	6.24	12.75
358	Cutting or forming holes for pipes 55 to 110 mm nominal size in common brickwork 328 mm thick, making good	nr	1.12	6.90	8.72	16.74

© NSR 01 Aug 2008 - 31 Jul 2009

	P : BUILDING FABRIC SUNDRIES		Mat. £	Lab. £	Plant £	Total £
P3150						
360	Cutting or forming holes for pipes 55 to 110 mm nominal size in facing brickwork 102 mm thick, making good	nr	1.12	5.82	6.59	13.53
362	Cutting or forming holes for pipes 55 to 110 mm nominal size in facing brickwork 215 mm thick, making good	nr	1.12	6.90	7.92	15.94
364	Cutting or forming holes for pipes 55 to 110 mm nominal size in facing brickwork 328 mm thick, making good	nr	1.12	9.48	11.63	22.23
366	Cutting or forming holes for pipes 55 to 110 mm nominal size in blockwork 60 mm thick, making good	nr	0.75	1.51	1.68	3.94
368	Cutting or forming holes for pipes 55 to 110 mm nominal size in blockwork 75 mm thick, making good	nr	0.75	2.59	2.71	6.05
370	Cutting or forming holes for pipes 55 to 110 mm nominal size in blockwork 100 mm thick, making good	nr	0.75	2.59	2.91	6.25
372	Cutting or forming holes for pipes 55 to 110 mm nominal size in blockwork 150 mm thick, making good	nr	0.75	2.80	3.23	6.78
374	Cutting or forming holes for pipes 55 to 110 mm nominal size in blockwork 200 mm thick, making good	nr	0.75	3.23	3.64	7.62
376	Cutting or forming holes for pipes 55 to 110 mm nominal size in blockwork 215 mm thick	nr	0.75	3.23	3.74	7.72
400	Cutting or forming holes for pipes over 110 mm nominal size in concrete 100 mm thick, making good	nr	0.93	7.54	8.40	16.87
402	Cutting or forming holes for pipes over 110 mm nominal size in concrete 200 mm thick, making good	nr	1.68	9.48	11.43	22.59
404	Cutting or forming holes for pipes over 110 mm nominal size in common brickwork 102 mm thick, making good	nr	1.68	6.03	6.72	14.43
406	Cutting or forming holes for pipes over 110 mm nominal size in common brickwork 215 mm thick, making good	nr	1.68	7.33	8.27	17.28
408	Cutting or forming holes for pipes over 110 mm nominal size in common brickwork 328 mm thick, making good	nr	1.68	9.05	10.98	21.71
410	Cutting or forming holes for pipes over 110 mm nominal size in facing brickwork 102 mm thick, making good	nr	1.68	7.97	8.85	18.50
412	Cutting or forming holes for pipes over 110 mm nominal size in facing brickwork 215 mm thick, making good	nr	1.68	9.48	10.63	21.79
414	Cutting or forming holes for pipes over 110 mm nominal size in facing brickwork 328 mm thick, making good	nr	1.68	12.28	14.56	28.52
416	Cutting or forming holes for pipes over 110 mm nominal size in blockwork 60 mm thick, making good	nr	1.12	2.15	2.36	5.63
418	Cutting or forming holes for pipes over 110 mm nominal size in blockwork 75 mm thick, making good	nr	1.12	3.23	3.49	7.84
420	Cutting or forming holes for pipes over 110 mm nominal size in blockwork 100 mm thick, making good	nr	1.12	3.45	3.81	8.38
422	Cutting or forming holes for pipes over 110 mm nominal size in blockwork 150 mm thick, making good	nr	1.12	3.66	4.14	8.92

© NSR 01 Aug 2008 - 31 Jul 2009

	P : BUILDING FABRIC SUNDRIES		Mat. £	Lab. £	Plant £	Total £
P3150						
424	Cutting or forming holes for pipes over 110 mm nominal size in blockwork 200 mm thick, making good	nr	1.12	4.31	4.76	10.19
426	Cutting or forming holes for pipes over 110 mm nominal size in blockwork 215 mm thick, making good	nr	1.12	4.53	4.99	10.64
450	Cutting or forming chases for services 1 nr 19 mm dia in concrete, making good	m	0.19	5.55	0.96	6.70
452	Cutting or forming chases for services 1 nr 19 mm dia in brickwork, making good	m	0.19	4.31	1.86	6.36
454	Cutting or forming chases for services 1 nr 19 mm dia in blockwork, making good	m	0.19	2.15	0.93	3.27
500	Cutting or forming chases for services 1 nr 25 mm dia in concrete, making good	m	0.37	7.40	1.28	9.05
502	Cutting or forming chases for services 1 nr 25 mm dia in brickwork, making good	m	0.37	5.17	1.96	7.50
504	Cutting or forming chases for services 1 nr 25 mm dia in blockwork, making good	m	0.37	2.59	0.98	3.94
506	Cutting or forming chases for services 1 nr 50 mm dia in concrete, making good	m	0.75	13.87	2.40	17.02
508	Cutting or forming chases for services 1 nr 50 mm dia in brickwork, making good	m	0.75	7.33	2.21	10.29
510	Cutting or forming chases for services 1 nr 50 mm dia in blockwork, making good	m	0.75	3.88	1.13	5.76
550	Cutting or forming chases for services 1 nr 50 x 50 mm in concrete, making good	m	0.75	13.87	2.40	17.02
552	Cutting or forming chases for services 1 nr 50 x 50 mm in brickwork, making good	m	0.75	6.46	2.11	9.32
554	Cutting or forming chases for services 1 nr 50 x 50 mm in blockwork, making good	m	0.75	3.45	1.08	5.28
600	Cutting or forming chases for services 2 nr 19 mm dia in concrete, making good	m	0.56	9.71	1.68	11.95
602	Cutting or forming chases for services 2 nr 19 mm dia in brickwork, making good	m	0.56	7.76	2.25	10.57
604	Cutting or forming chases for services 2 nr 19 mm dia in blockwork, making good	m	0.56	3.88	1.13	5.57
606	Cutting or forming chases for services 1 nr 100 x 100 mm in concrete, making good	m	1.86	31.44	5.44	38.74
608	Cutting or forming chases for services 1 nr 100 x 100 mm in brickwork, making good	m	1.86	12.07	2.75	16.68
610	Cutting or forming chases for services 1 nr 100 x 100 mm in blockwork, making good	m	1.86	7.54	1.55	10.95
625	Cutting or forming chases for services 2 nr 25 mm dia in concrete, making good	m	0.75	13.41	2.32	16.48

© NSR 01 Aug 2008 - 31 Jul 2009

	P : BUILDING FABRIC SUNDRIES		Mat. £	Lab. £	Plant £	Total £
P3150						
627	Cutting or forming chases for services 2 nr 25 mm dia in brickwork, making good	m	0.75	9.27	2.43	12.45
629	Cutting or forming chases for services 2 nr 25 mm dia in blockwork, making good	m	0.75	4.53	1.20	6.48
650	Cutting or forming chases for services 2 nr 50 mm dia in concrete, making good	m	1.86	24.97	4.32	31.15
652	Cutting or forming chases for services 2 nr 50 mm dia in brickwork, making good	m	1.86	11.64	2.70	16.20
654	Cutting or forming chases for services 2 nr 50 mm dia in blockwork, making good	m	1.86	6.90	1.48	10.24
656	Cutting or forming chases for services 1 nr 150 x 100 mm in concrete, making good	m	3.73	37.91	6.56	48.20
658	Cutting or forming chases for services 1 nr 150 x 100 mm in brickwork, making good	m	3.73	16.38	2.57	22.68
660	Cutting or forming chases for services 1 nr 150 x 100 mm in blockwork, making good	m	3.73	10.78	1.92	16.43
P3160	**Labours in reinforced concrete where reinforcement cut (50% addition on chases and holes in unreinforced concrete)**					
300	Cutting or forming holes for pipes ne 55 mm nominal size in concrete 100 mm thick, making good	nr	0.37	5.82	6.84	13.03
302	Cutting or forming holes for pipes ne 55 mm nominal size in concrete 200 mm thick, making good	nr	0.37	7.11	8.45	15.93
350	Cutting or forming holes for pipes 55 to 110 mm nominal size in concrete 100 mm thick, making good	nr	0.75	7.97	9.10	17.82
352	Cutting or forming holes for pipes 55 to 110 mm nominal size in concrete 200 mm thick, making good	nr	1.12	10.34	12.33	23.79
400	Cutting or forming holes for pipes over 110 mm nominal size in concrete 100 mm thick, making good	nr	0.93	11.21	12.23	24.37
402	Cutting or forming holes for pipes over 110 mm nominal size in concrete 200 mm thick, making good	nr	1.68	14.22	17.14	33.04
450	Cutting or forming chases for services 1 nr 19 mm dia in concrete, making good	m	0.19	8.32	2.82	11.33
500	Cutting or forming chases for services 1 nr 25 mm dia in concrete, making good	m	0.37	11.10	3.35	14.82
506	Cutting or forming chases for services 1 nr 50 mm dia in concrete, making good	m	0.75	20.80	5.02	26.57
550	Cutting or forming chases for services 1 nr 50 x 50 mm in concrete, making good	m	0.75	20.80	5.02	26.57
600	Cutting or forming chases for services 2 nr 19 mm dia in concrete, making good	m	0.56	14.33	3.88	18.77
606	Cutting or forming chases for services 1 nr 100 x 100 mm in concrete, making good	m	1.86	47.16	9.65	58.67

© NSR 01 Aug 2008 - 31 Jul 2009   P4

	P : BUILDING FABRIC SUNDRIES		Mat. £	Lab. £	Plant £	Total £
P3160						
625	Cutting or forming chases for services 2 nr 25 mm dia in concrete, making good	m	0.75	19.88	4.86	25.49
650	Cutting or forming chases for services 2 nr 50 mm dia in concrete, making good	m	1.86	37.45	7.94	47.25
656	Cutting or forming chases for services 1 nr 150 x 100 mm in concrete, making good	m	3.73	56.87	11.36	71.96
	P32 : Holes/chases/covers/supports for electrical services					
P3270	**General builders work**					
800	Cutting or forming holes, mortices, sinkings and chases, concealed service, luminaire point, making good	nr	0.49	25.17		25.66
802	Cutting or forming holes, mortices, sinkings and chases, concealed service, socket outlet point, making good	nr	0.43	18.13		18.56
804	Cutting or forming holes, mortices, sinkings and chases, concealed service, fitting outlet point, making good	nr	0.43	18.13		18.56
806	Cutting or forming holes, mortices, sinkings and chases, concealed service, equipment and control gear point, making good	nr	0.59	30.21		30.80
850	Cutting or forming holes, mortices, sinkings and chases, exposed service, luminaire point, making good	nr	0.17	14.10		14.27
852	Cutting or forming holes, mortices, sinkings and chases, exposed service, socket outlet point, making good	nr	0.14	9.06		9.20
854	Cutting or forming holes, mortices, sinkings and chases, exposed service, fitting outlet point, making good	nr	0.14	9.06		9.20
856	Cutting or forming holes, mortices, sinkings and chases, exposed service, equipment and control gear point, making good	nr	0.20	15.10		15.30

© NSR 01 Aug 2008 - 31 Jul 2009

P5

P : BUILDING FABRIC SUNDRIES	Mat. £	Lab. £	Plant £	Total £

			Mat. £	Lab. £	Plant £	Total £
	R : PLUMBING : WORKS OF ALTERATION/SMALL WORKS/REPAIR					
	R10 : SOIL AND WASTE					
R1001	**RAINWATER GOODS : supply only plastic rainwater goods with fittings**					
002	100 mm gutter	m	5.81			5.81
004	110 mm gutter	m	6.42			6.42
006	131 mm gutter	m	7.62			7.62
008	150 mm gutter	m	15.99			15.99
010	112 mm square gutter	m	12.07			12.07
012	68 mm downpipe	m	6.22			6.22
014	110 mm downpipe	m	18.92			18.92
016	65 mm square downpipe	m	6.88			6.88
R1005	**RAINWATER GOODS : fix only plastic rainwater goods with fittings**					
002	100 mm gutter	m		13.31		13.31
004	110 mm gutter	m		13.31		13.31
006	131 mm gutter	m		14.47		14.47
008	150 mm gutter	m		16.21		16.21
010	112 mm square gutter	m		13.31		13.31
012	68 mm downpipe	m		8.68		8.68
014	110 mm downpipe	m		11.00		11.00
016	65 mm square downpipe	m		8.68		8.68
R1010	**RAINWATER GOODS : remove damaged plastic rainwater goods, fix new rainwater goods with fittings**					
002	100 mm gutter	m		20.26		20.26
004	110 mm gutter	m		20.26		20.26
006	131 mm gutter	m		22.00		22.00
008	150 mm gutter	m		24.31		24.31
010	112 mm square gutter	m		20.26		20.26
012	68 mm downpipe	m		13.31		13.31
014	110 mm downpipe	m		16.79		16.79
016	65 mm square downpipe	m		13.31		13.31

© NSR 01 Aug 2008 - 31 Jul 2009

R1

	R : PLUMBING : WORKS OF ALTERATION/SMALL WORKS/REPAIR		Mat. £	Lab. £	Plant £	Total £
R1015	RAINWATER GOODS : remove damaged, supply and fix new plastic rainwater goods with fittings					
002	100 mm gutter	m	5.81	20.26		26.07
004	110 mm gutter	m	6.42	20.26		26.68
006	131 mm gutter	m	7.62	22.00		29.62
008	150 mm gutter	m	15.99	24.31		40.30
010	112 mm square gutter	m	12.07	20.26		32.33
012	68 mm downpipe	m	6.22	13.31		19.53
014	110 mm downpipe	m	18.92	16.79		35.71
016	65 mm square downpipe	m	6.88	13.31		20.19
R1020	RAINWATER GOODS : supply and fix plastic rainwater goods with fittings					
002	100 mm gutter	m	5.81	13.31		19.12
004	110 mm gutter	m	6.42	13.31		19.73
006	131 mm gutter	m	7.62	14.47		22.09
008	150 mm gutter	m	15.99	16.21		32.20
010	112 mm square gutter	m	12.07	13.31		25.38
012	68 mm downpipe	m	6.22	8.68		14.90
014	110 mm downpipe	m	18.92	11.00		29.92
016	65 mm square downpipe	m	6.88	8.68		15.56
R1025	SOIL AND WASTE : supply only plastic soil and waste goods and fittings					
000	82 mm soil	m	17.75			17.75
002	110 mm soil	m	19.50			19.50
004	160 mm soil	m	52.79			52.79
006	32 mm waste, welded	m	4.16			4.16
008	40 mm waste, welded	m	4.75			4.75
010	50 mm waste, welded	m	7.68			7.68
012	32 mm waste, weld and seal ring	m	5.37			5.37
014	40 mm waste, weld and seal ring	m	6.55			6.55
016	50 mm waste, weld and seal ring	m	9.36			9.36
018	32 mm bottle 'P' trap	nr	8.15			8.15
020	40 mm bottle 'P' trap	nr	9.72			9.72
022	50 mm 'P' trap	nr	17.64			17.64

© NSR 01 Aug 2008 - 31 Jul 2009

R2

R : PLUMBING : WORKS OF ALTERATION/SMALL WORKS/REPAIR				Mat. £	Lab. £	Plant £	Total £
R1025							
	024	90 mm 'P' trap	nr	52.00	4.69		56.69
	026	32 mm 'S' trap	nr	9.27			9.27
	028	40 mm 'S' trap	nr	10.84			10.84
R1030		SOIL AND WASTE : fix only plastic soil and waste goods and fittings					
	000	82 mm soil	m		14.47		14.47
	002	110 mm soil	m		14.47		14.47
	004	160 mm soil	m		15.63		15.63
	006	32 mm waste, welded	m		4.35		4.35
	008	40 mm waste, welded	m		4.35		4.35
	010	50 mm waste, welded	m		4.69		4.69
	012	32 mm waste, weld and seal ring	m		3.68		3.68
	014	40 mm waste, weld and seal ring	m		3.68		3.68
	016	50 mm waste, weld and seal ring	m		4.02		4.02
	018	32 mm bottle 'P' trap	nr		3.35		3.35
	020	40 mm bottle 'P' trap	nr		3.35		3.35
	022	50 mm 'P' trap	nr		4.02		4.02
	024	90 mm 'P' trap	nr		4.69		4.69
	026	32 mm 'S' trap	nr		3.35		3.35
	028	40 mm 'S' trap	nr		3.68		3.68
R1035		SOIL AND WASTE : remove damaged, fix only new plastic soil and waste goods and fittings					
	000	82 mm soil	m		20.26		20.26
	002	110 mm soil	m		20.26		20.26
	004	160 mm soil	m		23.73		23.73
	006	32 mm waste, welded	m		6.69		6.69
	008	40 mm waste, welded	m		6.69		6.69
	010	50 mm waste, welded	m		7.03		7.03
	012	32 mm waste, weld and seal ring	m		5.69		5.69
	014	40 mm waste, weld and seal ring	m		5.69		5.69
	016	50 mm waste, weld and seal ring	m		6.02		6.02
	018	32 mm bottle 'P' trap	m		5.02		5.02

© NSR 01 Aug 2008 - 31 Jul 2009   R3

R : PLUMBING : WORKS OF ALTERATION/SMALL WORKS/REPAIR			Mat. £	Lab. £	Plant £	Total £
R1035						
020	40 mm bottle 'P' trap	nr		5.02		5.02
022	50 mm 'P' trap	nr		6.02		6.02
024	90 mm 'P' trap	nr		7.03		7.03
026	32 mm 'S' trap	nr		5.02		5.02
028	40 mm 'S' trap	nr		5.69		5.69
R1040	**SOIL AND WASTE : supply and fix plastic soil and waste goods and fittings**					
000	82 mm soil	m	17.75	14.47		32.22
002	110 mm soil	m	19.50	14.47		33.97
004	160 mm soil	m	52.79	15.63		68.42
006	32 mm waste, welded	m	4.16	4.35		8.51
008	40 mm waste, welded	m	4.75	4.35		9.10
010	50 mm waste, welded	m	7.68	4.69		12.37
012	32 mm waste, weld and seal ring	m	5.37	3.68		9.05
014	40 mm waste, weld and seal ring	m	6.55	3.68		10.23
016	50 mm waste, weld and seal ring	m	9.36	4.02		13.38
018	32 mm bottle 'P' trap	nr	8.15	3.35		11.50
020	40 mm bottle 'P' trap	nr	9.72	3.35		13.07
022	50 mm 'P' trap	nr	17.64	4.02		21.66
024	90 mm 'P' trap	nr	52.00	4.69		56.69
026	32 mm 'S' trap	nr	9.27	3.35		12.62
028	40 mm 'S' trap	nr	10.84	3.68		14.52
R1045	**SOIL AND WASTE : remove damaged, supply and fix new plastic soil and waste goods and fittings**					
000	82 mm soil	m	17.75	20.26		38.01
002	110 mm soil	m	19.50	20.26		39.76
004	160 mm soil	m	52.79	23.73		76.52
006	32 mm waste, welded	m	4.16	6.69		10.85
008	40 mm waste, welded	m	4.75	6.69		11.44
010	50 mm waste, welded	m	7.68	7.03		14.71
012	32 mm waste, weld and seal ring	m	5.37	5.69		11.06
014	40 mm waste, weld and seal ring	m	6.55	5.69		12.24

© NSR 01 Aug 2008 - 31 Jul 2009   R4

R : PLUMBING : WORKS OF ALTERATION/SMALL WORKS/REPAIR			Mat. £	Lab. £	Plant £	Total £
R1045						
016	50 mm waste, weld and seal ring	m	9.36	6.02		15.38
018	32 mm bottle 'P' trap	nr	8.15	5.02		13.17
020	40 mm bottle 'P' trap	nr	9.72	5.02		14.74
022	50 mm 'P' trap	nr	17.64	6.02		23.66
024	90 mm 'P' trap	nr	52.00	7.03		59.03
026	32 mm 'S' trap	nr	9.27	5.02		14.29
028	40 mm 'S' trap	nr	10.84	5.69		16.53

© NSR 01 Aug 2008 - 31 Jul 2009   R5

R : PLUMBING : WORKS OF ALTERATION/SMALL WORKS/REPAIR		Mat. £	Lab. £	Plant £	Total £

© NSR 01 Aug 2008 - 31 Jul 2009   R6

			Mat. £	Lab. £	Plant £	Total £
	S : MECHANICAL SERVICES - PIPE SUPPLY SYSTEMS : WORKS OF ALTERATION/SMALL WORKS/REPAIR					
	S15 : HOT AND COLD WATER SERVICES : PLANT AND EQUIPMENT					
S1500	CYLINDERS : fix only indirect cylinders, BS1566 : Part 1 : 1972, copper, immersion heater, insulated					
010	900 x 400 mm, 11 in immersion	nr		26.05		26.05
012	1050 x 400 mm, 11 in immersion	nr		26.05		26.05
014	1050 x 400 mm, 27 in immersion	nr		34.73		34.73
016	1200 x 400 mm, 27 in immersion	nr		34.73		34.73
S1503	CYLINDERS : supply only indirect cylinders, BS1566 : Part 1 : 1972, copper, immersion heater, insulated					
010	900 x 400 mm, 11 in. immersion	nr	192.08			192.08
012	1050 x 400 mm, 11in. immersion	nr	186.86			186.86
014	1050 x 400 mm, 27in. Immersion	nr	193.16			193.16
016	1200 x 400 mm, 27 in. immersion	nr	251.18			251.18
S1506	CYLINDERS : supply and fix indirect cylinders, BS1566 : Part 1 : 1972, copper, immersion heater, insulated					
010	900 x 400 mm, 11 in immersion	nr	192.08	26.05		218.13
012	1050 x 400 mm, 11 in immersion	nr	186.86	26.05		212.91
014	1050 x 400 mm, 27 in immersion	nr	193.16	34.73		227.89
016	1200 x 400 mm, 27 in immersion	nr	251.18	34.73		285.91
S1509	CYLINDERS : fix only combination indirect cylinders, BS3198 : 1981, copper, cylindrical, immersion heater, insulated					
020	115 Litres hot, 25 litres cold capacity	nr		52.10		52.10
022	115 Litres hot, 45 litres cold capacity	nr		52.10		52.10
024	150 Litres hot, 45 litres cold capacity	nr		60.78		60.78
S1512	CYLINDERS : supply only combination indirect cylinders, BS3198 : 1981, copper, cylindrical, immersion heater, insulated					
020	900 x 450 mm, 11 in immersion	nr	279.76			279.76
022	1050 x 450 mm, 11 in immersion	nr	291.94			291.94
024	1200 x 450 mm, 27 in immersion	nr	332.27			332.27

© NSR 01 Aug 2008 - 31 Jul 2009   S1

S : MECHANICAL SERVICES - PIPE SUPPLY SYSTEMS : WORKS OF ALTERATION/SMALL WORKS/REPAIR			Mat. £	Lab. £	Plant £	Total £
S1515	CYLINDERS : supply and fix combination indirect cylinders, BS3198 : 1981, copper, cylindrical, immersion heater, insulated					
020	900 x 450 mm, 11 in immersion	nr	279.76	52.10		331.86
022	1050 x 450 mm, 11 in immersion	nr	291.94	52.10		344.04
024	1200 x 450 mm, 27 in immersion	nr	332.27	60.78		393.05
S1518	CIRCULATING PUMPS : fix only hot water service circulating pumps, bronze, in line glandless					
050	20 - 50 mm Diameter	nr		35.14		35.14
S1521	CIRCULATING PUMPS : supply only hot water circulating pumps, bronze, in line glandless					
050	20 - 50 mm Diameter	nr	409.00			409.00
S1524	CIRCULATING PUMPS : supply and fix hot water service circulating pumps, bronze, in line glandless					
050	20 - 50 mm diameter	nr	409.00	35.14		444.14
S1527	SHOWERS AND MIXING VALVES					
010	Remove damaged shower spray head, fix only new head	nr		14.06		14.06
011	Remove, supply and fix new shower spray head	nr	34.84	14.06		48.90
012	Supply and fix new shower spray head	nr	34.84	7.03		41.87
014	Remove damaged rigid shower assembly, fix only new assembly	nr		19.08		19.08
015	Remove, supply and fix new rigid shower assembly	nr	64.00	27.44		91.44
016	Supply and fix new rigid shower assembly	nr	64.00	19.08		83.08
018	Remove damaged flexible shower assembly and spray head, fix only new assembly	nr		17.40		17.40
019	Remove, supply and fix new flexible shower assembly and spray head	nr	71.00	19.08		90.08
020	Supply and fix new flexible shower assembly and spray head	nr	71.00	14.06		85.06
021	Remove damaged shower mixing valve control wheel / lever, fix only new wheel / lever	nr		14.06		14.06
022	Remove, supply and fix new shower mixing valve control wheel / lever	nr	28.00	14.06		42.06
024	Supply and fix new shower mixing valve control wheel / lever	nr	28.00	7.03		35.03
026	Dismantle, service manual shower mixing valve and re-assemble	nr		17.40		17.40
028	Dismantle, service strainer valve and re-assemble	nr		17.40		17.40
030	Remove damaged manual shower mixing valve, fix only new assembly	nr		14.06		14.06
031	Remove, supply and fix new manual shower mixing valve	nr	354.22	14.06		368.28
032	Supply and fix new manual shower mixing valve	nr	354.22	7.03		361.25

© NSR 01 Aug 2008 - 31 Jul 2009

S2

S : MECHANICAL SERVICES - PIPE SUPPLY SYSTEMS : WORKS OF ALTERATION/SMALL WORKS/REPAIR			Mat. £	Lab. £	Plant £	Total £
S1527						
034	Dismantle, service thermostatic shower mixing valve and re-assemble	nr		17.40		17.40
036	Remove damaged thermostatic shower mixing valve cartridge assembly, fix only new assembly	nr		14.06		14.06
037	Remove, supply and fix new thermostatic shower mixing valve cartridge assembly	nr	65.33	14.06		79.39
038	Supply and fix new thermostatic shower mixing valve cartridge assembly	nr	65.33	7.03		72.36
S1530	SHOWERS AND MIXING VALVES : fix only shower mixing valves, shower assembly, spray head, isolating/ check valves					
060	Concealed pattern	nr		118.81		118.81
062	Exposed pattern	nr		100.40		100.40
S1533	SHOWERS AND MIXING VALVES : supply only shower mixing valves, shower assembly, spray head, isolating/ check valves					
060	Concealed pattern	nr	354.23			354.23
062	Exposed pattern	nr	307.00			307.00
S1536	SHOWERS AND MIXING VALVES : supply and fix shower mixing valves, shower assembly, spray head, isolating/ check valves					
060	Concealed pattern	nr	354.23	118.81		473.04
062	Exposed pattern	nr	307.00	100.40		407.40
S1539	SHOWERS AND MIXING VALVES : fix only thermostatic mixing valves, including isolating/ check valves					
066	20 mm Diameter	nr		62.58		62.58
068	25 mm Diameter	nr		70.28		70.28
S1542	SHOWERS AND MIXING VALVES : supply only thermostatic mixing valves, including isolating/ check valves					
066	20 mm Diameter	nr	271.40			271.40
068	25 mm Diameter	nr	280.55			280.55
S1545	SHOWERS AND MIXING VALVES : supply and fix thermostatic mixing valves, including isolating/ check valves					
066	20 mm diameter	nr	271.40	62.58		333.98
068	25 mm diameter	nr	280.55	70.28		350.83

© NSR 01 Aug 2008 - 31 Jul 2009   S3

S : MECHANICAL SERVICES - PIPE SUPPLY SYSTEMS : WORKS OF ALTERATION/SMALL WORKS/REPAIR				Mat. £	Lab. £	Plant £	Total £
S1548	STORAGE TANKS : fix only cold water storage tank, BS4213, polypropylene, rectangular, cover, ballvalve, insulated						
092	18 Litre capacity		nr		15.73		15.73
100	68 Litre capacity		nr		26.05		26.05
102	91 Litre capacity		nr		26.05		26.05
104	114 Litre capacity		nr		34.73		34.73
106	227 Litre capacity		nr		52.10		52.10
S1551	STORAGE TANKS : supply only cold water storage tank, BS4213, polyproylene, rectangular, cover, ballvalve, insulated						
092	18 Litre capacity		nr	72.74			72.74
100	68 Litre capacity		nr	72.74			72.74
102	91 Litre capacity		nr	75.74			75.74
104	114 Litre capacity		nr	90.32			90.32
106	227 Litre capacity		nr	155.53			155.53
S1554	STORAGE TANKS : supply and fix cold water storage tank, BS4213, polypropylene, rectangular, cover, ballvalve, insulated						
092	18 litre capacity		nr	72.74	15.73		88.47
100	68 litre capacity		nr	72.74	26.05		98.79
102	91 litre capacity		nr	75.74	26.05		101.79
104	114 litre capacity		nr	90.32	34.73		125.05
106	227 litre capacity		nr	155.53	52.10		207.63
S1557	STORAGE TANKS : fix only cold water storage tank, BS417, galvanised, rectangular, cover, ballvalve, insulated						
100	45 Litre capacity		nr		34.73		34.73
102	91 Litre capacity		nr		43.41		43.41
104	114 Litre capacity		nr		48.04		48.04
106	227 Litre capacity		nr		74.09		74.09
S1560	STORAGE TANKS : supply only cold water storage tank, BS417, galvanised, rectangular, cover, ballvalve, insulated						
100	45 Litre capacity		nr	94.81			94.81
102	91 Litre capacity		nr	121.48			121.48
104	114 Litre capacity		nr	134.52			134.52
106	227 Litre capacity		nr	190.14			190.14

© NSR 01 Aug 2008 - 31 Jul 2009   S4

S : MECHANICAL SERVICES - PIPE SUPPLY SYSTEMS : WORKS OF ALTERATION/SMALL WORKS/REPAIR			Mat. £	Lab. £	Plant £	Total £
S1563	STORAGE TANKS : supply and fix cold water storage tank, BS417, galvanised, rectangular, cover, ballvalve, insulated					
100	45 litre capacity	nr	94.81	34.73		129.54
102	91 litre capacity	nr	121.48	43.41		164.89
104	114 litre capacity	nr	134.52	48.04		182.56
106	227 litre capacity	nr	190.14	74.09		264.23
S1566	STORAGE TANKS : supply only new ball valve and float					
400	15 mm, Portsmouth pattern, high pressure, plastic float	nr	11.26			11.26
402	22 mm, Portsmouth pattern, high pressure, plastic float	nr	14.13			14.13
430	15 mm, equilibrium pattern, high pressure, plastic float	nr	11.26			11.26
432	22 mm, equilibrium pattern, high pressure, plastic float	nr	15.15			15.15
434	28 mm, equilibrium pattern, high pressure, plastic float	nr	59.40			59.40
436	50 mm, equilibrium pattern, high pressure, flanged connections, bronze	nr	306.54			306.54
438	75 mm, equilibrium pattern, high pressure, flanged connections, bronze	nr	1188.74			1188.74
440	75 mm, equilibrium pattern, high pressure, flanged connections, cast iron	nr	357.57			357.57
S1569	STORAGE TANKS : remove damaged ball valve and float, fix only new ball valve and float					
400	15 mm, Portsmouth pattern, high pressure, plastic float	nr		20.08		20.08
402	22 mm, Portsmouth pattern, high pressure, plastic float	nr		23.43		23.43
430	15 mm, equilibrium pattern, high pressure, plastic float	nr		20.08		20.08
432	22 mm, equilibrium pattern, high pressure, plastic float	nr		23.43		23.43
434	28 mm, equilibrium pattern, high pressure, plastic float	nr		26.77		26.77
436	50 mm, equilibrium pattern, high pressure, flanged connections, bronze	nr		40.16		40.16
438	75 mm, equilibrium pattern, high pressure, flanged connections, bronze	nr		53.55		53.55
440	75 mm, equilibrium pattern, high pressure, flanged connections, cast iron	nr		53.55		53.55
S1572	STORAGE TANKS: remove damaged ball valve and float, supply and fix new ball valve and float					
400	15 mm, Portsmouth pattern, high pressure, plastic float	nr	6.11	20.08		26.19
402	22 mm, Portsmouth pattern, high pressure, plastic float	nr	12.89	23.43		36.32
430	15 mm, equilibrium pattern, high pressure, plastic float	nr	19.35	20.08		39.43
432	22 mm, equilibrium pattern, high pressure, plastic float	nr	29.73	23.43		53.16

© NSR 01 Aug 2008 - 31 Jul 2009 S5

S : MECHANICAL SERVICES - PIPE SUPPLY SYSTEMS : WORKS OF ALTERATION/SMALL WORKS/REPAIR			Mat. £	Lab. £	Plant £	Total £
S1572						
434	28 mm, equilibrium pattern, high pressure, plastic float	nr	59.40	26.77		86.17
436	50 mm, equilibrium pattern, high pressure, flanged connections, bronze	nr	306.54	40.16		346.70
438	75 mm, equilibrium pattern, high pressure, flanged connections, bronze	nr	1188.74	53.55		1242.29
440	75 mm, equilibrium pattern, high pressure, flanged connections, cast iron	nr	357.57	53.55		411.12
S1575	LABORATORY FITTINGS : remove damaged laboratory tap, single and multiple type, fix only new tap					
900	Bench mounted pattern	nr		20.08		20.08
902	Wall mounted pattern	nr		15.06		15.06
904	Remote control mounted pattern	nr		21.75		21.75
S1578	LABORATORY FITTINGS : supply only laboratory tap					
900	Bench mounted pattern	nr	48.68			48.68
902	Wall mounted pattern	nr	102.12			102.12
904	Remote control mounted pattern	nr	89.39			89.39
S1581	LABORATORY FITTINGS : remove damaged laboratory tap, single and multiple type, fix only new tap					
900	Bench mounted pattern	nr		20.08		20.08
902	Wall mounted pattern	nr		15.06		15.06
904	Remote control mounted pattern	nr		21.75		21.75
	S17 : HOT AND COLD WATER SERVICES : PIPEWORK AND ANCILLARIES					
S1700	PIPEWORK : repair leaking pipe or fitting not exceeding 1000 mm long					
044	20 - 25 mm Diameter, polyethylene	nr		10.04		10.04
046	32 - 50 mm Diameter, polyethylene	nr		12.72		12.72
048	63 mm Diameter, polyethylene	nr		16.73		16.73
050	15 - 22 mm Diameter, copper	nr		9.37		9.37
052	28 - 35 mm Diameter, copper	nr		14.39		14.39
054	42 - 54 mm Diameter, copper	nr		20.08		20.08
056	67 mm Diameter, copper	nr		24.89		24.89
058	76 mm Diameter, copper	nr		30.68		30.68

© NSR 01 Aug 2008 - 31 Jul 2009   S6

S : MECHANICAL SERVICES - PIPE SUPPLY SYSTEMS : WORKS OF ALTERATION/SMALL WORKS/REPAIR			Mat. £	Lab. £	Plant £	Total £
S1700						
060	108 mm Diameter, copper	nr		46.31		46.31
062	15 - 20 mm Diameter, galvanised mild steel	nr		11.04		11.04
064	25 - 32 mm Diameter, galvanised mild steel	nr		17.07		17.07
066	40 - 50 mm Diameter, galvanised mild steel	nr		24.10		24.10
068	65 - 80 mm Diameter, galvanised mild steel	nr		28.94		28.94
070	100 mm Diameter, galvanised mild steel	nr		34.73		34.73
072	15 mm Diameter, lead	nr		10.04		10.04
074	20 mm Diameter, lead	nr		11.04		11.04
076	25 mm Diameter, lead	nr		12.05		12.05
S1703	PIPEWORK : supply and fit MDPE pipe, BS6572, compression fittings, BS864 : Part 3					
050	20 mm Diameter	m	1.09	6.02		7.11
052	extra over for 20 mm fittings with one end	nr	3.51	2.68		6.19
054	extra over for 20 mm fittings with two ends	nr	3.22	3.35		6.57
056	extra over for 20 mm fittings with three ends	nr	5.79	5.35		11.14
100	25 mm Diameter	m	1.23	6.69		7.92
102	extra over for 25 mm fittings with one end	nr	4.71	3.01		7.72
104	extra over for 25 mm fittings with two ends	nr	3.39	3.68		7.07
106	extra over for 25 mm fittings with three ends	nr	9.06	5.69		14.75
150	32 mm Diameter	m	2.08	7.70		9.78
152	extra over for 32 mm fittings with one end	nr	6.28	3.68		9.96
154	extra over for 32 mm fittings with two ends	nr	8.04	4.35		12.39
156	extra over for 32 mm fittings with three ends	nr	11.35	6.69		18.04
200	50 mm Diameter	m	5.06	8.37		13.43
202	extra over for 50 mm fittings with one end	nr	15.17	5.02		20.19
204	extra over for 50 mm fittings with two ends	nr	18.52	6.02		24.54
206	extra over for 50 mm fittings with three ends	nr	26.47	8.70		35.17
250	63 mm Diameter	m	8.05	9.37		17.42
252	extra over for 63 mm fittings with one end	nr	21.48	6.02		27.50
254	extra over for 63 mm fittings with two ends	nr	27.87	6.69		34.56
256	extra over for 63 mm fittings with three ends	nr	41.00	9.37		50.37

© NSR 01 Aug 2008 - 31 Jul 2009   S7

	S : MECHANICAL SERVICES - PIPE SUPPLY SYSTEMS : WORKS OF ALTERATION/SMALL WORKS/REPAIR		Mat. £	Lab. £	Plant £	Total £
S1706	PIPEWORK : remove damaged pipework and fittings, supply and fix new, not exceeding 1000 long					
050	15 - 22 mm Diameter, copper	nr	19.74	27.21		46.95
052	28 - 35 mm Diameter, copper	nr	90.89	39.36		130.25
054	42 - 54 mm Diameter, copper	nr	164.04	54.41		218.45
056	67 mm Diameter, copper	nr	556.96	65.99		622.95
058	76 mm Diameter, copper	nr	832.19	77.56		909.75
060	108 mm Diameter, copper	nr	1332.67	93.77		1426.44
062	15 - 20 mm Diameter, galvanised mild steel	nr	25.15	28.94		54.09
064	25 - 32 mm Diameter, galvanised mild steel	nr	49.06	36.47		85.53
066	40 - 50 mm Diameter, galvanised mild steel	nr	82.08	54.99		137.07
068	65 - 80 mm Diameter, galvanised mild steel	nr	252.99	95.51		348.50
070	100 mm Diameter, galvanised mild steel	nr	468.42	118.08		586.50
S1709	PIPEWORK : supply and fit MDPE pipe, BS6572, fusion fittings, BS864 : Part 3					
050	20 mm Diameter	m	1.09	6.02	1.89	9.00
052	extra over for 20 mm fittings with one end	nr	11.88	2.01	0.63	14.52
054	extra over for 20 mm fittings with two ends	nr	7.28	3.68	1.16	12.12
056	extra over for 20 mm fittings with three ends	nr	15.10	5.35	1.68	22.13
100	25 mm Diameter	m	1.23	6.69	2.10	10.02
102	extra over for 25 mm fittings with one end	nr	11.88	2.01	0.63	14.52
104	extra over for 25 mm fittings with two ends	nr	7.28	3.68	1.16	12.12
106	extra over for 25 mm fittings with three ends	nr	15.10	5.35	1.68	22.13
150	32 mm Diameter	m	2.08	7.70	2.42	12.20
152	extra over for 32 mm fittings with one end	nr	11.88	2.68	0.84	15.40
154	extra over for 32 mm fittings with two ends	nr	7.28	4.35	1.37	13.00
156	extra over for 32 mm fittings with three ends	nr	17.16	6.02	1.89	25.07
200	50 mm Diameter	m	5.06	8.37	2.63	16.06
202	extra over for 50 mm fittings with one end	nr	19.91	3.35	1.05	24.31
204	extra over for 50 mm fittings with two ends	nr	12.08	4.69	1.47	18.24
206	extra over for 50 mm fittings with three ends	nr	26.51	6.69	2.10	35.30
250	63 mm Diameter	m	8.05	9.37	2.94	20.36
252	extra over for 63 mm fittings with one end	nr	22.02	3.35	1.05	26.42
254	extra over for 63 mm fittings with two ends	nr	13.31	4.69	1.47	19.47
256	extra over for 63 mm fittings with three ends	nr	37.75	7.36	2.31	47.42

© NSR 01 Aug 2008 - 31 Jul 2009   S8

S : MECHANICAL SERVICES - PIPE SUPPLY SYSTEMS : WORKS OF ALTERATION/SMALL WORKS/REPAIR			Mat. £	Lab. £	Plant £	Total £
S1709						
300	90 mm Diameter	m	15.78	10.71	3.37	29.86
302	extra over for 90mm end cap	nr	39.12	4.02	1.26	44.40
304	extra over for 90mm coupling	nr	19.98	6.69	2.10	28.77
306	extra over for 90mm tee	nr	61.42	10.04	3.16	74.62
S1712	PIPEWORK : remove damaged final connections serving appliances, laboratory bench fittings and the like, supply and fix new pipework and fittings, including setting as necessary, not exceeding 1000 mm long					
080	15 mm Diameter copper, hot water	nr	11.82	40.16		51.98
082	15 mm Diameter copper, cold water	nr	11.82	40.16		51.98
084	22 mm Diameter copper, hot water	nr	21.26	41.83		63.09
086	22 mm Diameter copper, cold water	nr	21.26	41.83		63.09
088	15 mm Diameter galvanised mild steel, hot water	nr	22.06	60.78		82.84
090	15 mm Diameter galvanised mild steel, cold water	nr	22.06	60.78		82.84
092	20 mm Diameter galvanised mild steel, hot water	nr	25.15	63.67		88.82
094	20 mm Diameter galvanised mild steel, cold water	nr	25.15	63.67		88.82
S1715	PIPEWORK : supply and fit PVC pipe, BS3505 : 1986, solvent welded fittings, BS4346 : Part 1 : 1969					
100	10 mm Diameter	m	1.35	4.02		5.37
102	extra over for 10mm fittings with one end	nr	0.91	0.67		1.58
104	extra over for 10mm fittings with two ends	nr	1.05	1.34		2.39
106	extra over for 10mm fittings with three ends	nr	5.67	2.01		7.68
150	12 mm Diameter	m	1.47	4.02		5.49
152	extra over for 12mm fittings with one end	nr	0.91	0.67		1.58
154	extra over for 12mm fittings with two ends	nr	1.05	1.34		2.39
156	extra over for 12mm fittings with three ends	nr	5.67	2.01		7.68
200	16 mm Diameter	m	1.50	4.02		5.52
202	extra over for 16mm fittings with one end	nr	1.02	0.67		1.69
204	extra over for 16mm fittings with two ends	nr	2.91	1.34		4.25
206	extra over for 16mm fittings with three ends	nr	6.18	2.01		8.19
250	20 mm Diameter	m	2.21	4.35		6.56
252	extra over for 20mm fittings with one end	nr	1.02	0.67		1.69
254	extra over for 20mm fittings with two ends	nr	3.35	1.34		4.69

© NSR 01 Aug 2008 - 31 Jul 2009   S9

	S : MECHANICAL SERVICES - PIPE SUPPLY SYSTEMS : WORKS OF ALTERATION/SMALL WORKS/REPAIR			Mat. £	Lab. £	Plant £	Total £
S1715							
256		extra over for 20mm fittings with three ends	nr	8.88	2.01		10.89
300	25 mm Diameter		m	2.87	5.02		7.89
302		extra over for 25mm fittings with one end	nr	1.47	0.67		2.14
304		extra over for 25mm fittings with two ends	nr	3.55	1.34		4.89
306		extra over for 25mm fittings with three ends	nr	9.50	2.01		11.51
350	32 mm Diameter		m	4.25	5.69		9.94
352		extra over for 32mm fittings with one end	nr	1.63	1.00		2.63
354		extra over for 32mm fittings with two ends	nr	4.32	1.67		5.99
356		extra over for 32mm fittings with three ends	nr	10.68	2.68		13.36
400	40 mm Diameter		m	6.29	6.02		12.31
402		extra over for 40mm fittings with one end	nr	2.57	1.00		3.57
404		extra over for 40mm fittings with two ends	nr	6.18	1.67		7.85
406		extra over for 40mm fittings with three ends	nr	13.06	2.68		15.74
450	50 mm Diameter		m	9.16	6.69		15.85
452		extra over for 50mm fittings with one end	nr	4.32	1.00		5.32
454		extra over for 50mm fittings with two ends	nr	7.77	1.67		9.44
456		extra over for 50mm fittings with three ends	nr	16.44	2.68		19.12
S1718	PIPEWORK : supply and fit light gauge copper pipes, BS2871 : Part 1 : 1971 : Table X, compression fittings, BS864 : Part 2						
200	15 mm Diameter		m	12.01	6.69		18.70
202		extra over for 15mm fittings with one end	nr	2.89	2.01		4.90
204		extra over for 15mm fittings with two ends	nr	2.23	3.01		5.24
206		extra over for 15mm fittings with three ends	nr	3.23	4.69		7.92
250	22 mm Diameter		m	13.44	8.03		21.47
252		extra over for 22mm fittings with one end	nr	3.46	2.68		6.14
254		extra over for 22mm fittings with two ends	nr	3.76	3.35		7.11
256		extra over for 22mm fittings with three ends	nr	5.28	5.02		10.30
300	28 mm Diameter		m	16.95	8.70		25.65
302		extra over for 28mm fittings with one end	nr	11.23	3.01		14.24
304		extra over for 28mm fittings with two ends	nr	13.49	3.68		17.17
306		extra over for 28mm fittings with three ends	nr	24.55	5.69		30.24
350	35 mm Diameter		m	39.37	10.04		49.41
352		extra over for 35mm fittings with one end	nr	20.27	3.68		23.95

© NSR 01 Aug 2008 - 31 Jul 2009   S10

S : MECHANICAL SERVICES - PIPE SUPPLY SYSTEMS : WORKS OF ALTERATION/SMALL WORKS/REPAIR

				Mat. £	Lab. £	Plant £	Total £
S1718							
	354	extra over for 35mm fittings with two ends	nr	33.94	4.35		38.29
	356	extra over for 35mm fittings with three ends	nr	44.87	6.69		51.56
	400	42 mm Diameter	m	47.88	11.71		59.59
	402	extra over for 42mm fittings with one end	nr	33.14	4.02		37.16
	404	extra over for 42mm fittings with two ends	nr	47.51	4.69		52.20
	406	extra over for 42mm fittings with three ends	nr	69.05	7.36		76.41
	450	54 mm Diameter	m	61.60	13.39		74.99
	452	extra over for 54mm fittings with one end	nr	46.22	5.02		51.24
	454	extra over for 54mm fittings with two ends	nr	80.70	5.69		86.39
	456	extra over for 54mm fittings with three ends	nr	111.12	8.37		119.49
S1721		PIPEWORK : supply and fix light gauge copper pipes (TX), EN1057 R250, capillary fittings, BS864 : Part 2					
	200	15 mm Diameter	m	6.72	6.69		13.41
	202	extra over for fittings with one end	nr	2.08	2.01		4.09
	204	extra over for fittings with two ends	nr	0.60	3.01		3.61
	206	extra over for fittings with three ends	nr	1.13	4.69		5.82
	250	22 mm Diameter	m	13.44	8.03		21.47
	252	extra over for fittings with one end	nr	3.88	2.68		6.56
	254	extra over for fittings with two ends	nr	1.56	3.68		5.24
	256	extra over for fittings with three ends	nr	3.61	6.02		9.63
	300	28 mm Diameter	m	16.95	8.70		25.65
	302	extra over for fittings with one end	nr	6.92	3.01		9.93
	304	extra over for fittings with two ends	nr	3.08	4.35		7.43
	306	extra over for fittings with three ends	nr	8.55	6.69		15.24
	350	35 mm Diameter	m	39.37	10.04		49.41
	352	extra over for fittings with one end	nr	15.30	4.02		19.32
	354	extra over for fittings with two ends	nr	13.39	5.35		18.74
	356	extra over for fittings with three ends	nr	21.80	9.04		30.84
	400	42 mm Diameter	m	47.88	11.71		59.59
	402	extra over for fittings with one end	nr	26.34	4.35		30.69
	404	extra over for fittings with two ends	nr	22.13	6.36		28.49
	406	extra over for fittings with three ends	nr	34.98	9.37		44.35
	450	54 mm Diameter	m	61.60	13.39		74.99

© NSR 01 Aug 2008 - 31 Jul 2009   S11

S : MECHANICAL SERVICES - PIPE SUPPLY SYSTEMS : WORKS OF ALTERATION/SMALL WORKS/REPAIR				Mat. £	Lab. £	Plant £	Total £
S1721							
452		extra over for fittings with one end	nr	36.77	4.69		41.46
454		extra over for fittings with two ends	nr	45.71	7.03		52.74
456		extra over for fittings with three ends	nr	70.53	10.04		80.57
S1724	PIPEWORK : supply and fit galvanised mild steel pipes, BS1387 : 1985, screwed fittings, BS143						
200	15 mm Diameter		m	5.22	8.37		13.59
202		extra over for fittings with one end	nr	10.02	5.02		15.04
204		extra over for fittings with two ends	nr	3.49	10.04		13.53
206		extra over for fittings with three ends	nr	28.63	14.06		42.69
250	20 mm Diameter		m	5.45	10.04		15.49
252		extra over for fittings with one end	nr	13.28	7.36		20.64
254		extra over for fittings with two ends	nr	4.03	13.39		17.42
256		extra over for fittings with three ends	nr	35.21	18.74		53.95
300	25 mm Diameter		m	7.61	11.71		19.32
302		extra over for fittings with one end	nr	21.32	8.37		29.69
304		extra over for fittings with two ends	nr	5.19	15.06		20.25
306		extra over for fittings with three ends	nr	56.74	21.08		77.82
350	32 mm Diameter		m	8.65	13.39		22.04
352		extra over for fittings with one end	nr	29.34	10.04		39.38
354		extra over for fittings with two ends	nr	7.63	17.74		25.37
356		extra over for fittings with three ends	nr	103.38	25.10		128.48
400	40 mm Diameter		m	10.95	15.06		26.01
402		extra over for fittings with one end	nr	36.26	10.71		46.97
404		extra over for fittings with two ends	nr	9.77	20.08		29.85
406		extra over for fittings with three ends	nr	103.38	28.11		131.49
450	50 mm Diameter		m	15.36	18.41		33.77
452		extra over for fittings with one end	nr	50.07	12.72		62.79
454		extra over for fittings with two ends	nr	14.65	24.10		38.75
456		extra over for fittings with three ends	nr	151.24	33.47		184.71
S1727	PIPEWORK : supply and fit short water connections to appliances, including pipework, fittings, setting as necessary, jointing to distribution pipework, not exceeding 1000 mm long						
330	10 mm Diameter uPVC		nr	5.54	9.37		14.91

© NSR 01 Aug 2008 - 31 Jul 2009   S12

S : MECHANICAL SERVICES - PIPE SUPPLY SYSTEMS : WORKS OF ALTERATION/SMALL WORKS/REPAIR			Mat. £	Lab. £	Plant £	Total £
S1727						
332	15 mm Diameter uPVC	nr	9.15	9.37		18.52
334	20 mm Diameter uPVC	nr	12.05	10.04		22.09
336	15 mm Diameter copper	nr	10.36	30.12		40.48
338	22 mm Diameter copper	nr	49.82	35.47		85.29
340	15 mm Diameter galvanised mild steel	nr	19.92	32.13		52.05
342	20 mm Diameter galvanised mild steel	nr	39.19	35.14		74.33
S1730	VALVES AND ANCILLARIES : remove damaged hand wheel or lockshield isolating valve, supply and fix new					
370	15 mm Diameter	nr	25.29	7.36		32.65
372	20 mm Diameter	nr	34.46	8.70		43.16
374	25 mm Diameter	nr	52.64	10.71		63.35
376	32 mm Diameter	nr	74.29	12.38		86.67
378	40 mm Diameter	nr	92.25	14.73		106.98
380	50 mm Diameter	nr	146.14	28.94		175.08
382	65 mm Diameter	nr	371.41	34.73		406.14
384	80 mm Diameter	nr	538.20	46.89		585.09
386	100 mm Diameter	nr	476.29	52.10		528.39
S1733	VALVES AND ANCILLARIES : supply and fit isolating valves wheel-head and lockshield					
400	15 mm Diameter	nr	25.29	7.36		32.65
402	20 mm Diameter	nr	34.46	8.70		43.16
404	25 mm Diameter	nr	52.64	10.71		63.35
406	32 mm Diameter	nr	74.29	12.38		86.67
408	40 mm Diameter	nr	92.25	14.73		106.98
410	50 mm Diameter	nr	146.14	16.73		162.87
S1736	VALVES AND ANCILLARIES : remove damaged check valve, supply and fix new					
370	15 mm Diameter	nr	51.58	7.36		58.94
372	20 mm Diameter	nr	65.90	8.70		74.60
374	25 mm Diameter	nr	90.39	10.71		101.10
376	32 mm Diameter	nr	142.57	12.38		154.95
378	40 mm Diameter	nr	168.66	14.73		183.39

© NSR 01 Aug 2008 - 31 Jul 2009   S13

S : MECHANICAL SERVICES - PIPE SUPPLY SYSTEMS : WORKS OF ALTERATION/SMALL WORKS/REPAIR				Mat. £	Lab. £	Plant £	Total £
S1736							
380	50 mm Diameter		nr	251.39	28.94		280.33
382	65 mm Diameter		nr	240.62	34.73		275.35
384	80 mm Diameter		nr	467.06	46.89		513.95
386	100 mm Diameter		nr	616.47	52.10		668.57
S1739	VALVES AND ANCILLARIES : supply and fit check valves						
400	15 mm Diameter		nr	28.52	7.36		35.88
402	20 mm Diameter		nr	42.49	8.70		51.19
404	25 mm Diameter		nr	60.42	10.71		71.13
406	32 mm Diameter		nr	81.28	12.38		93.66
408	40 mm Diameter		nr	95.03	14.73		109.76
410	50 mm Diameter		nr	153.34	16.73		170.07
S1742	VALVES AND ANCILLARIES : supply and fit strainers						
400	15 mm Diameter		nr	39.66	7.36		47.02
402	20 mm Diameter		nr	52.06	8.70		60.76
404	25 mm Diameter		nr	79.03	10.71		89.74
406	32 mm Diameter		nr	118.44	12.38		130.82
408	40 mm Diameter		nr	144.16	14.73		158.89
410	50 mm Diameter		nr	265.55	16.73		282.28
S1745	VALVES AND ANCILLARIES : tighten or repack valve glands, or re-washer stopcock						
350	15 - 50 mm Diameter		nr	26.78	14.06		40.84
352	65 - 100 mm Diameter		nr	26.78	21.42		48.20
S1748	VALVES AND ANCILLARIES : remove damaged stopcock, supply and fix new						
358	15 mm Diameter		nr	16.45	14.73		31.18
360	22 mm Diameter		nr	17.74	17.40		35.14
362	25 mm Diameter		nr	34.58	21.42		56.00
364	32 mm Diameter		nr	63.60	24.77		88.37
366	40 mm Diameter		nr	74.19	29.45		103.64
368	50 mm Diameter		nr	92.59	33.47		126.06

© NSR 01 Aug 2008 - 31 Jul 2009   S14

S : MECHANICAL SERVICES - PIPE SUPPLY SYSTEMS : WORKS OF ALTERATION/SMALL WORKS/REPAIR			Mat. £	Lab. £	Plant £	Total £
S1751	**VALVES AND ANCILLARIES : supply and fit stopcocks**					
420	15 mm Diameter	nr	16.45	7.36		23.81
422	22 mm Diameter	nr	17.74	8.70		26.44
424	28 mm Diameter	nr	34.58	10.71		45.29
426	35 mm Diameter	nr	63.60	12.38		75.98
428	42 mm Diameter	nr	74.19	14.73		88.92
430	54 mm Diameter	nr	92.59	16.73		109.32
S1754	**VALVES AND ANCILLARIES : remove damaged double regulating valve, supply and fix new**					
370	15 mm Diameter	nr	56.34	7.36		63.70
372	20 mm Diameter	nr	124.18	8.70		132.88
374	25 mm Diameter	nr	147.68	10.71		158.39
376	32 mm Diameter	nr	198.01	12.38		210.39
378	40 mm Diameter	nr	287.15	14.73		301.88
380	50 mm Diameter	nr	410.75	28.94		439.69
382	65 mm Diameter	nr	302.58	39.94		342.52
384	80 mm Diameter	nr	317.43	46.89		364.32
386	100 mm Diameter	nr	412.08	52.10		464.18
S1757	**VALVES AND ANCILLARIES : supply and fit ballofix valves**					
420	15 mm Diameter	nr	11.10	10.04		21.14
422	22 mm Diameter	nr	19.85	11.71		31.56
S1760	**VALVES AND ANCILLARIES : remove damaged draincock, supply and fix new**					
370	15 mm Diameter	nr	35.66	10.71		46.37
372	20 mm Diameter	nr	51.99	15.39		67.38
S1763	**VALVES AND ANCILLARIES : supply and fit draincocks**					
432	15 mm Diameter, hose outlet	nr	44.96	10.04		55.00
434	20 mm Diameter, hose outlet	nr	60.91	11.71		72.62
436	15 mm Diameter, hose union	nr	25.94	10.04		35.98
438	20 mm Diameter, hose union	nr	32.78	11.71		44.49
S1766	**VALVES AND ANCILLARIES : supply and fit bib taps and backplate elbow**					
400	15 mm Diameter	nr	5.76	10.04		15.80

© NSR 01 Aug 2008 - 31 Jul 2009   S15

S : MECHANICAL SERVICES - PIPE SUPPLY SYSTEMS : WORKS OF ALTERATION/SMALL WORKS/REPAIR			Mat. £	Lab. £	Plant £	Total £
S1766						
402	20 mm Diameter	nr	11.46	11.71		23.17
S1769	**VALVES AND ANCILLARIES : supply and fit manual air vents**					
440	8 mm Diameter	nr	8.30	3.35		11.65
442	15 mm diameter	nr	12.90	6.36		19.26
444	22 mm diameter	nr	17.03	7.36		24.39
S1772	**VALVES AND ANCILLARIES : supply and fit venting**					
455	15 mm diameter air vents including drain pipe assembly	nr	123.12	10.04		133.16
460	22 mm diameter air vents including drain pipe assembly	nr	111.60	21.08		132.68
465	32 mm diameter air vents including drain pipe assembly	nr	135.25	24.77		160.02
470	50 mm Diameter air bottle assemblies including drain pipe assembly	nr	135.25	35.14		170.39
S1775	**VALVES AND ANCILLARIES : supply and fit test plugs**					
460	12 mm Diameter	nr	10.97	15.06		26.03
S1778	**VALVES AND ANCILLARIES : supply and fit instruments**					
472	100 mm Diameter dial pressure gauges, syphons and cocks	nr	137.43	15.06		152.49
474	150 mm Diameter dial pressure gauges, syphons and cocks	nr	133.89	15.06		148.95
476	100 mm Diameter dial thermometers including pockets	nr	165.50	15.06		180.56
478	150 mm Diameter dial thermometers including pockets	nr	186.25	15.06		201.31
480	100 mm Diameter straight stem thermometers including pockets	nr	91.84	15.06		106.90
482	100 mm Diameter angle stem thermometers including pockets	nr	94.76	15.06		109.82
S1781	**VALVES AND ANCILLARIES : supply and fit automatic flush valve**					
498	15 mm Diameter	nr	162.50	10.04		172.54
S1784	**GENERALLY**					
950	Draining down domestic hot water system, including cylinder and tank, refilling	nr		40.16		40.16
952	Draining down semi-commercial property hot water system, including cylinder and tank, refilling	nr		80.32		80.32
960	Draining down cold water system, cleaning out 60 gallon cistern, refilling	nr		45.18		45.18
962	Draining down cold water system, cleaning out 100 gallon cistern, refilling	nr		50.20		50.20

© NSR 01 Aug 2008 - 31 Jul 2009   S16

S : MECHANICAL SERVICES - PIPE SUPPLY SYSTEMS : WORKS OF ALTERATION/SMALL WORKS/REPAIR			Mat. £	Lab. £	Plant £	Total £
S1784						
970	Draining down cold water system, cleaning out 500 gallon cistern, refilling	nr		75.30		75.30
	S25 : SPACE HEATING SERVICES : PLANT AND EQUIPMENT					
S2500	FEED AND EXPANSION TANKS : fix only feed and expansion tanks, BS4213, polypropylene, rectangular, cover, ballvalve, insulated					
900	18 Litre capacity	nr		15.73		15.73
902	68 Litre capacity	nr		26.05		26.05
S2515	FEED AND EXPANSION TANKS : supply only feed and expansion tanks, BS4213, polypropylene, rectangular, cover, ballvalve, insulated					
900	18 Litre capacity	nr	72.74			72.74
902	68 Litre capacity	nr	143.77			143.77
S2530	FEED AND EXPANSION TANKS : supply and fix feed and expansion tanks, BS4213, polypropylene, rectangular, cover, ballvalve, insulated					
900	18 litre capacity	nr	72.74	15.73		88.47
902	68 litre capacity	nr	143.77	26.05		169.82
S2545	FEED AND EXPANSION TANKS : fix only feed and expansion tanks, BS417, galvanised, rectangular, cover, ballvalve, insulated					
900	18 Litre capacity	nr		40.16		40.16
902	68 Litre capacity	nr		60.24		60.24
S2560	FEED AND EXPANSION TANKS : supply only feed and expansion tanks, BS417, galvanised, rectangular, cover, ballvalve, insulated					
900	18 Litre capacity	nr	109.22			109.22
902	68 Litre capacity	nr	146.31			146.31
S2575	FEED AND EXPANSION TANKS : supply and fix feed and expansion tanks, BS417, galvanised, rectangular, cover, ballvalve, insulated					
900	18 litre capacity	nr	109.22	40.16		149.38
902	68 litre capacity	nr	146.31	60.24		206.55

© NSR 01 Aug 2008 - 31 Jul 2009   S17

S : MECHANICAL SERVICES - PIPE SUPPLY SYSTEMS : WORKS OF ALTERATION/SMALL WORKS/REPAIR			Mat. £	Lab. £	Plant £	Total £
	S27 : SPACE HEATING SERVICES : PIPEWORK AND ANCILLARIES					
S2700	PIPEWORK : repair leaking pipe or fitting not exceeding 1000 mm long					
200	15 - 22 mm Diameter, copper	nr		9.37		9.37
202	28 - 35 mm Diameter, copper	nr		14.39		14.39
204	42 - 54 mm Diameter, copper	nr		20.08		20.08
206	67 mm Diameter, copper	nr		28.78		28.78
208	76 mm Diameter, copper	nr		35.47		35.47
210	108 mm Diameter, copper	nr		53.55		53.55
212	15 - 20 mm Diameter, mild steel	nr		8.06		8.06
214	25 - 32 mm Diameter, mild steel	nr		12.45		12.45
216	40 - 50 mm Diameter, mild steel	nr		17.58		17.58
218	65 mm Diameter, mild steel	nr		20.75		20.75
220	80 mm Diameter, mild steel	nr		25.64		25.64
222	100 mm Diameter, mild steel	nr		29.30		29.30
S2702	PIPEWORK : supply and fix light gauge copper pipes (TX), EN1057 R250, capillary fittings, BS864 : Part 2					
200	15 mm Diameter	m	6.72	6.69		13.41
202	extra over for fittings with one end	nr	2.08	2.01		4.09
204	extra over for fittings with two ends	nr	0.60	3.01		3.61
206	extra over for fittings with three ends	nr	1.13	4.69		5.82
250	22 mm Diameter	m	13.44	8.03		21.47
252	extra over for fittings with one end	nr	3.88	2.68		6.56
254	extra over for fittings with two ends	nr	1.56	3.68		5.24
256	extra over for fittings with three ends	nr	3.61	6.02		9.63
300	28 mm Diameter	m	16.95	8.70		25.65
302	extra over for fittings with one end	nr	6.92	3.01		9.93
304	extra over for fittings with two ends	nr	3.08	4.35		7.43
306	extra over for fittings with three ends	nr	8.55	6.69		15.24
350	35 mm Diameter	m	39.37	10.04		49.41
352	extra over for fittings with one end	nr	15.30	4.02		19.32
354	extra over for fittings with two ends	nr	13.39	5.35		18.74
356	extra over for fittings with three ends	nr	21.80	9.04		30.84
400	42 mm Diameter	m	47.88	11.71		59.59

© NSR 01 Aug 2008 - 31 Jul 2009   S18

S : MECHANICAL SERVICES - PIPE SUPPLY SYSTEMS : WORKS OF ALTERATION/SMALL WORKS/REPAIR

			Mat. £	Lab. £	Plant £	Total £
S2702						
402	extra over for fittings with one end	nr	26.34	4.35		30.69
404	extra over for fittings with two ends	nr	22.13	6.36		28.49
406	extra over for fittings with three ends	nr	34.98	9.37		44.35
450	54 mm Diameter	m	61.60	13.39		74.99
452	extra over for fittings with one end	nr	36.77	4.69		41.46
454	extra over for fittings with two ends	nr	45.71	7.03		52.74
456	extra over for fittings with three ends	nr	70.53	10.04		80.57
S2704	PIPEWORK : remove damaged pipework and fittings, supply and fix new, not exceeding 1000 mm long					
198	15 mm diameter, copper	nr	10.36	22.57		32.93
200	22 mm Diameter, copper	nr	19.74	27.21		46.95
202	28 mm Diameter, copper	nr	24.25	39.36		63.61
203	42 mm diameter, copper	nr	68.76	43.41		112.17
204	54 mm Diameter, copper	nr	100.10	54.41		154.51
206	67 mm Diameter, copper	nr	556.96	65.99		622.95
208	76 mm Diameter, copper	nr	832.19	77.56		909.75
210	108 mm Diameter, copper	nr	1332.67	93.77		1426.44
212	20 mm Diameter, mild steel	nr	65.89	30.68		96.57
214	32 mm Diameter, mild steel	nr	137.39	39.94		177.33
216	50 mm Diameter, mild steel	nr	200.95	60.78		261.73
218	65 mm Diameter, mild steel	nr	169.39	95.51		264.90
220	80 mm Diameter, mild steel	nr	134.71	118.08		252.79
222	100 mm Diameter, mild steel	nr	549.85	149.34		699.19
S2706	PIPEWORK : supply and fix black mild steel pipes, BS1387 : 1985, screwed fittings, BS143					
200	15 mm Diameter	m	4.77	8.37		13.14
202	Extra over for fittings with one end	nr	8.16	5.02		13.18
204	Extra over for fittings with two ends	nr	2.58	10.04		12.62
206	Extra over for fittings with three ends	nr	21.21	14.06		35.27
250	20 mm Diameter	m	4.77	10.04		14.81
252	Extra over for fittings with one end	nr	2.97	7.36		10.33
254	Extra over for fittings with two ends	nr	2.97	13.39		16.36

© NSR 01 Aug 2008 - 31 Jul 2009

S19

	S : MECHANICAL SERVICES - PIPE SUPPLY SYSTEMS : WORKS OF ALTERATION/SMALL WORKS/REPAIR		Mat. £	Lab. £	Plant £	Total £
S2706						
256	Extra over for fittings with three ends	nr	26.08	18.74		44.82
300	25 mm Diameter	m	8.65	11.71		20.36
302	Extra over for fittings with one end	nr	15.79	8.37		24.16
304	Extra over for fittings with two ends	nr	3.84	15.06		18.90
306	Extra over for fittings with three ends	nr	42.03	21.08		63.11
350	32 mm Diameter	m	8.65	13.39		22.04
352	Extra over for fittings with one end	nr	23.45	10.04		33.49
354	Extra over for fittings with two ends	nr	5.65	17.74		23.39
356	Extra over for fittings with three ends	nr	79.45	25.10		104.55
400	40 mm Diameter	m	14.01	15.06		29.07
402	Extra over for fittings with one end	nr	26.86	10.71		37.57
404	Extra over for fittings with two ends	nr	7.23	20.08		27.31
406	Extra over for fittings with three ends	nr	79.45	28.11		107.56
450	50 mm Diameter	m	14.01	18.41		32.42
452	Extra over for fittings with one end	nr	39.49	12.72		52.21
454	Extra over for fittings with two ends	nr	10.85	24.10		34.95
456	Extra over for fittings with three ends	nr	112.03	33.47		145.50
500	65 mm Diameter	m	19.05	19.08		38.13
502	Extra over for Cap 16W	nr	82.68	15.39		98.07
504	Extra over for Socket 14W	nr	18.59	29.45		48.04
506	Extra over for Female Tee 12W	nr	409.49	41.83		451.32
550	80 mm Diameter	m	24.25	24.43		48.68
552	Extra over for Cap 16W	nr	123.87	18.07		141.94
554	Extra over for Socket 14W	nr	25.99	34.81		60.80
556	Extra over for Female Tee 12W	nr	444.29	48.86		493.15
600	100 mm Diameter	m	33.85	33.13		66.98
602	Extra over for Cap 16W	nr	215.04	24.10		239.14
604	Extra over for Socket 14W	nr	49.61	48.19		97.80
606	Extra over for Female Tee 12W	nr	713.34	67.60		780.94
S2708	PIPEWORK : supply and fix black mild steel pipes, BS1387 : 1985, weldable fittings, BS1965					
200	15 mm Diameter	m	4.19	4.35		8.54
202	Extra over for fittings with one end	nr	43.56	6.02		49.58

© NSR 01 Aug 2008 - 31 Jul 2009   S20

S : MECHANICAL SERVICES - PIPE SUPPLY SYSTEMS : WORKS OF ALTERATION/SMALL WORKS/REPAIR				Mat. £	Lab. £	Plant £	Total £
S2708							
204		Extra over for fittings with two ends	nr	9.40	12.05		21.45
206		Extra over for fittings with three ends	nr	88.90	18.07		106.97
250	20 mm Diameter		m	4.77	5.02		9.79
252		Extra over for fittings with one end	nr	43.56	8.03		51.59
254		Extra over for fittings with two ends	nr	9.20	16.06		25.26
256		Extra over for fittings with three ends	nr	88.90	24.10		113.00
300	25 mm Diameter		m	6.97	5.69		12.66
302		Extra over for fittings with one end	nr	43.56	10.04		53.60
304		Extra over for fittings with two ends	nr	12.22	20.08		32.30
306		Extra over for fittings with three ends	nr	88.90	30.12		119.02
350	32 mm Diameter		m	8.65	6.69		15.34
352		Extra over for fittings with one end	nr	43.56	12.72		56.28
354		Extra over for fittings with two ends	nr	14.58	25.43		40.01
356		Extra over for fittings with three ends	nr	88.90	38.15		127.05
400	40 mm Diameter		m	10.08	7.70		17.78
402		Extra over for fittings with one end	nr	43.82	16.06		59.88
404		Extra over for fittings with two ends	nr	14.74	32.13		46.87
406		Extra over for fittings with three ends	nr	88.90	48.19		137.09
450	50 mm Diameter		m	14.01	8.37		22.38
452		Extra over for fittings with one end	nr	51.20	20.08		71.28
454		Extra over for fittings with two ends	nr	20.16	40.16		60.32
456		Extra over for fittings with three ends	nr	93.60	80.32		173.92
500	65 mm Diameter		m	19.05	10.04		29.09
502		Extra over for Cap	nr	60.14	26.10		86.24
504		Extra over for Elbow	nr	25.96	52.21		78.17
506		Extra over for Tee	nr	136.54	78.31		214.85
550	80 mm Diameter		m	24.25	10.71		34.96
552		Extra over for Cap	nr	61.20	31.12		92.32
554		Extra over for Elbow	nr	28.18	62.25		90.43
556		Extra over for Tee	nr	136.76	93.37		230.13
600	100 mm Diameter		m	33.85	11.71		45.56
602		Extra over for Cap	nr	80.04	40.16		120.20
604		Extra over for Elbow	nr	42.64	80.32		122.96

© NSR 01 Aug 2008 - 31 Jul 2009

S21

S : MECHANICAL SERVICES - PIPE SUPPLY SYSTEMS : WORKS OF ALTERATION/SMALL WORKS/REPAIR			Mat. £	Lab. £	Plant £	Total £
S2708						
606	Extra over for Tee	nr	170.16	120.48		290.64
S2710	PIPEWORK : remove radiator and existing valves, set aside for re-use, remove damaged radiator final connections, supply and fix pipework and fittings including setting as necessary, refix existing valves and radiator, not exceeding 1000 mm long					
300	15 mm Diameter pipework, copper	nr	10.36	121.56		131.92
302	22 mm Diameter pipework, copper	nr	19.74	131.40		151.14
304	28 mm Diameter pipework, copper	nr	27.71	140.08		167.79
306	15 mm Diameter pipework, mild steel	nr	54.35	129.66		184.01
308	20 mm Diameter pipework, mild steel	nr	65.89	166.13		232.02
310	25 mm Diameter pipework, mild steel	nr	80.47	182.33		262.80
S2712	PIPEWORK : remove fan assisted convector access plate, remove damaged final connections and set aside existing valves for re-use, supply and fix new pipework and fittings including setting as necessary, refix existing valves, refix casing panel, not exceeding 1000 mm long					
300	15 mm Diameter pipework, copper	nr	10.36	127.34		137.70
302	22 mm Diameter pipework, copper	nr	19.74	137.18		156.92
304	28 mm Diameter pipework, copper	nr	27.71	145.87		173.58
306	15 mm Diameter pipework, mild steel	nr	54.35	135.45		189.80
308	20 mm Diameter pipework, mild steel	nr	65.89	171.91		237.80
310	25 mm Diameter pipework, mild steel	nr	80.47	188.12		268.59
S2714	PIPEWORK : supply and fit short water connections to radiators, including pipework, fittings, isolating valves, setting as necessary, jointing to distribution pipework, not exceeding 1000 mm long					
350	15 mm Diameter, copper	nr	12.35	33.47		45.82
352	22 mm Diameter, copper	nr	23.61	36.81		60.42
354	28 mm Diameter, copper	nr	44.71	40.16		84.87
358	15 mm Diameter, black mild steel	nr	56.34	73.63		129.97
360	20 mm Diameter, black mild steel	nr	69.76	87.01		156.77
362	25 mm Diameter, black mild steel	nr	97.47	100.40		197.87

© NSR 01 Aug 2008 - 31 Jul 2009

S22

S : MECHANICAL SERVICES - PIPE SUPPLY SYSTEMS : WORKS OF ALTERATION/SMALL WORKS/REPAIR			Mat. £	Lab. £	Plant £	Total £
S2716	PIPEWORK : supply and fit short water connections to radiators, including pipework, fittings, air vent, isolating valve, setting as necessary, jointing to distribution pipework, not exceeding 1000 mm long					
350	15 mm Diameter, copper	nr	10.36	33.47		43.83
352	22 mm Diameter, copper	nr	19.74	36.81		56.55
354	28 mm Diameter, copper	nr	27.71	40.16		67.87
358	15 mm Diameter, black mild steel	nr	54.35	73.63		127.98
360	20 mm Diameter, black mild steel	nr	65.89	87.01		152.90
362	25 mm Diameter, black mild steel	nr	80.47	100.40		180.87
S2718	PIPEWORK : supply and fit short water connections to convector, including pipework, fittings, isolating valves, draincocks, setting as necessary, jointing to distribution pipework, not exceeding 1000 mm long					
350	15 mm Diameter, copper	nr	13.43	43.51		56.94
352	22 mm Diameter, copper	nr	24.69	46.85		71.54
354	28 mm Diameter, copper	nr	46.54	50.20		96.74
358	15 mm Diameter, black mild steel	nr	57.42	83.67		141.09
360	20 mm Diameter, black mild steel	nr	70.84	97.05		167.89
362	25 mm Diameter, black mild steel	nr	98.55	110.44		208.99
S2720	PIPEWORK : supply and fit short water connections to perimeter heating, including pipework, fittings, isolating valves, draincocks, air vents, jointing to distribution pipework, not exceeding 1000 mm long					
350	15 mm diameter, copper	nr	13.14	40.16		53.30
352	22 mm Diameter, copper	nr	24.39	46.85		71.24
354	28 mm Diameter, copper	nr	47.18	50.20		97.38
356	35 mm Diameter, copper	nr	206.91	53.55		260.46
362	25 mm Diameter, black mild steel	nr	135.02	110.44		245.46
364	32 mm Diameter, black mild steel	nr	246.96	123.83		370.79
S2722	PIPEWORK : supply and fit perimeter heating interconnecting pipework and fittings not exceeding 1000 mm long					
350	15 mm diameter, copper	nr	11.46	19.41		30.87
352	22 mm Diameter, copper	nr	19.74	21.42		41.16
354	28 mm Diameter, copper	nr	20.49	24.10		44.59
356	35 mm Diameter, copper	nr	52.05	27.78		79.83
362	25 mm Diameter, black mild steel	nr	104.39	25.10		129.49

© NSR 01 Aug 2008 - 31 Jul 2009

S23

S : MECHANICAL SERVICES - PIPE SUPPLY SYSTEMS : WORKS OF ALTERATION/SMALL WORKS/REPAIR			Mat. £	Lab. £	Plant £	Total £
S2722						
364	32 mm Diameter, black mild steel	nr	130.71	30.12		160.83
S2724	PIPEWORK : supply and fit short water connections to unit heater, including pipework, fittings, isolating valves, draincocks, air vents, jointing to distribution pipework, not exceeding 1000 mm long					
362	25 mm Diameter, black mild steel	nr	135.02	25.10		160.12
364	32 mm Diameter, black mild steel	nr	246.96	30.12		277.08
366	40 mm Diameter, black mild steel	nr	285.28	36.14		321.42
S2726	PIPEWORK : supply and fit short water connections to radiant panel, including pipework, fittings, isolating valves, draincocks, air vents, jointing to distribution pipework, not exceeding 1000 mm long					
358	15 mm Diameter, black mild steel	nr	66.50	22.09		88.59
360	20 mm Diameter, black mild steel	nr	99.76	24.10		123.86
362	25 mm Diameter, black mild steel	nr	135.02	25.10		160.12
364	32 mm Diameter, black mild steel	nr	246.96	30.12		277.08
S2728	PIPEWORK : remove perimeter heating front plate, remove damaged final connections and set aside existing valves for re-use, supply and fix new pipework and fittings including setting as necessary, refix existing valves, refix front plate, not exceeding 1000 mm long					
300	15 mm Diameter pipework, copper	nr	10.36	121.56		131.92
302	22 mm Diameter pipework, copper	nr	19.74	131.40		151.14
304	28 mm Diameter pipework, copper	nr	27.71	140.08		167.79
306	15 mm Diameter pipework, mild steel	nr	54.35	129.66		184.01
308	20 mm Diameter pipework, mild steel	nr	17.19	166.13		183.32
310	25 mm Diameter pipework, mild steel	nr	18.11	182.33		200.44
312	32 mm Diameter pipework, mild steel	nr	29.61	191.02		220.63
S2730	VALVES AND ANCILLARIES : remove damaged hand wheel or lockshield valve, supply and fix new, reset valve position					
620	15 mm Diameter	nr	14.36	12.73		27.09
622	20 mm Diameter	nr	16.96	15.05		32.01
624	25 mm Diameter	nr	23.42	18.52		41.94
626	32 mm Diameter	nr	35.09	21.42		56.51
628	40 mm Diameter	nr	49.42	25.47		74.89

© NSR 01 Aug 2008 - 31 Jul 2009   S24

S : MECHANICAL SERVICES - PIPE SUPPLY SYSTEMS : WORKS OF ALTERATION/SMALL WORKS/REPAIR			Mat. £	Lab. £	Plant £	Total £
S2730						
630	50 mm Diameter	nr	68.46	28.94		97.40
632	65 mm Diameter	nr	120.09	34.73		154.82
634	80 mm Diameter	nr	174.86	46.89		221.75
636	100 mm Diameter	nr	215.84	52.10		267.94
S2732	VALVES AND ANCILLARIES : remove damaged isolating valve, hand wheel or lockshield, supply and fix new					
610	15 - 25 mm Diameter	nr	47.25	11.71		58.96
612	32 - 50 mm Diameter	nr	131.85	17.07		148.92
S2734	VALVES AND ANCILLARIES : supply and fit isolating valve wheel-head and lockshield					
600	15 mm Diameter	nr	23.50	7.36		30.86
602	20 mm Diameter	nr	37.27	8.70		45.97
604	25 mm Diameter	nr	69.68	10.71		80.39
606	32 mm Diameter	nr	74.29	12.38		86.67
608	40 mm Diameter	nr	92.25	14.73		106.98
611	50 mm Diameter	nr	146.14	16.73		162.87
613	65 mm Diameter	nr	371.41	23.09		394.50
614	80 mm Diameter	nr	538.20	27.11		565.31
616	100 mm Diameter	nr	1427.06	30.12		1457.18
S2736	VALVES AND ANCILLARIES : remove damaged double regulating valve, supply and fix new					
620	15 mm Diameter	nr	56.34	12.73		69.07
622	20 mm Diameter	nr	96.55	15.05		111.60
624	25 mm Diameter	nr	121.13	18.52		139.65
626	32 mm Diameter	nr	169.60	21.42		191.02
628	40 mm Diameter	nr	244.05	25.47		269.52
630	50 mm Diameter	nr	357.09	21.42		378.51
632	65 mm Diameter	nr	511.83	39.94		551.77
634	80 mm Diameter	nr	626.48	46.89		673.37
636	100 mm Diameter	nr	859.74	52.10		911.84

© NSR 01 Aug 2008 - 31 Jul 2009

S25

S : MECHANICAL SERVICES - PIPE SUPPLY SYSTEMS : WORKS OF ALTERATION/SMALL WORKS/REPAIR			Mat. £	Lab. £	Plant £	Total £
S2738	VALVES AND ANCILLARIES : supply and fit double regulating valve					
600	15 mm Diameter	nr	56.34	7.36		63.70
602	20 mm Diameter	nr	96.55	8.70		105.25
604	25 mm Diameter	nr	121.13	10.71		131.84
606	32 mm Diameter	nr	169.60	12.38		181.98
608	40 mm Diameter	nr	244.05	14.73		258.78
610	50 mm Diameter	nr	357.09	16.73		373.82
612	65 mm Diameter	nr	511.83	23.09		534.92
614	80 mm Diameter	nr	626.48	27.11		653.59
616	100 mm Diameter	nr	859.74	30.12		889.86
S2740	VALVES AND ANCILLARIES : remove damaged metering station, supply and fix new					
620	15 mm Diameter	nr	15.34	17.37		32.71
622	20 mm Diameter	nr	21.96	20.26		42.22
624	25 mm Diameter	nr	24.47	22.57		47.04
626	32 mm Diameter	nr	27.93	26.05		53.98
628	40 mm Diameter	nr	33.10	29.52		62.62
630	50 mm Diameter	nr	43.49	31.26		74.75
632	65 mm Diameter	nr	108.91	39.94		148.85
634	80 mm Diameter	nr	125.03	46.89		171.92
636	100 mm Diameter	nr	149.94	52.10		202.04
S2742	VALVES AND ANCILLARIES : supply and fit metering station					
600	15 mm Diameter	nr	15.34	10.04		25.38
602	20 mm Diameter	nr	21.96	11.71		33.67
604	25 mm Diameter	nr	24.47	13.05		37.52
606	32 mm Diameter	nr	27.93	15.06		42.99
608	40 mm Diameter	nr	33.10	17.07		50.17
610	50 mm Diameter	nr	43.49	18.07		61.56
612	65 mm Diameter	nr	108.91	23.09		132.00
614	80 mm Diameter	nr	125.03	27.11		152.14
616	100 mm Diameter	nr	149.94	30.12		180.06

© NSR 01 Aug 2008 - 31 Jul 2009   S26

S : MECHANICAL SERVICES - PIPE SUPPLY SYSTEMS : WORKS OF ALTERATION/SMALL WORKS/REPAIR			Mat. £	Lab. £	Plant £	Total £
S2744	**VALVES AND ANCILLARIES : remove damaged orifice valve, supply and fix new**					
620	15 mm Diameter	nr	77.28	17.37		94.65
622	20 mm Diameter	nr	124.18	20.26		144.44
624	25 mm Diameter	nr	147.68	22.57		170.25
626	32 mm Diameter	nr	198.01	26.05		224.06
628	40 mm Diameter	nr	287.15	29.52		316.67
630	50 mm Diameter	nr	410.75	31.26		442.01
632	65 mm Diameter	nr	609.72	39.94		649.66
634	80 mm Diameter	nr	737.35	46.89		784.24
636	100 mm Diameter	nr	971.31	52.10		1023.41
S2746	**VALVES AND ANCILLARIES : supply and fit orifice valve**					
600	15 mm Diameter	nr	77.28	10.04		87.32
602	20 mm Diameter	nr	124.18	11.71		135.89
604	25 mm Diameter	nr	147.68	13.05		160.73
606	32 mm Diameter	nr	198.01	15.06		213.07
608	40 mm Diameter	nr	287.15	17.07		304.22
610	50 mm Diameter	nr	410.75	18.07		428.82
612	65 mm Diameter	nr	609.72	18.19		627.91
614	80 mm Diameter	nr	737.35	21.31		758.66
616	100 mm Diameter	nr	971.31	23.39		994.70
S2748	**VALVES AND ANCILLARIES : remove damaged commissioning set, supply and fix new**					
620	15 mm Diameter	nr	39.33	17.37		56.70
622	20 mm Diameter	nr	65.52	20.26		85.78
624	25 mm Diameter	nr	79.50	22.57		102.07
626	32 mm Diameter	nr	104.73	26.05		130.78
628	40 mm Diameter	nr	139.17	29.52		168.69
630	50 mm Diameter	nr	201.18	31.26		232.44
632	65 mm Diameter	nr	521.22	39.94		561.16
634	80 mm Diameter	nr	618.12	46.89		665.01
636	100 mm Diameter	nr	793.56	52.10		845.66

© NSR 01 Aug 2008 - 31 Jul 2009   S27

S : MECHANICAL SERVICES - PIPE SUPPLY SYSTEMS : WORKS OF ALTERATION/SMALL WORKS/REPAIR			Mat. £	Lab. £	Plant £	Total £
S2750	**VALVES AND ANCILLARIES : supply and fit commissioning set**					
600	15 mm Diameter	nr	39.33	10.04		49.37
602	20 mm Diameter	nr	65.52	11.71		77.23
604	25 mm Diameter	nr	79.50	13.05		92.55
606	32 mm Diameter	nr	104.73	15.06		119.79
608	40 mm Diameter	nr	139.17	17.07		156.24
610	50 mm Diameter	nr	201.18	18.07		219.25
612	65 mm Diameter	nr	521.22	23.09		544.31
614	80 mm Diameter	nr	618.12	27.11		645.23
616	100 mm Diameter	nr	793.56	30.12		823.68
S2752	**VALVES AND ANCILLARIES : supply and fit ball valve**					
600	15 mm Diameter	nr	9.95	10.04		19.99
602	20 mm Diameter	nr	12.49	11.71		24.20
604	25 mm Diameter	nr	18.28	13.05		31.33
606	32 mm Diameter	nr	28.03	12.38		40.41
608	40 mm Diameter	nr	40.67	16.40		57.07
610	50 mm Diameter	nr	63.31	16.73		80.04
612	65 mm Diameter	nr	144.20	18.19		162.39
614	80 mm Diameter	nr	226.65	21.31		247.96
616	100 mm Diameter	nr	473.92	23.39		497.31
S2754	**VALVES AND ANCILLARIES : remove check valve, supply and fix new**					
620	15 mm Diameter	nr	22.13	12.73		34.86
622	20 mm Diameter	nr	26.32	15.05		41.37
624	25 mm Diameter	nr	36.46	37.05		73.51
626	32 mm Diameter	nr	61.78	21.42		83.20
628	40 mm Diameter	nr	76.86	25.47		102.33
630	50 mm Diameter	nr	117.88	28.94		146.82
632	65 mm Diameter	nr	240.62	34.73		275.35
634	80 mm Diameter	nr	340.23	46.89		387.12
636	100 mm Diameter	nr	366.65	52.10		418.75
S2756	**VALVES AND ANCILLARIES : supply and fit check valve**					
600	15 mm Diameter	nr	22.13	10.04		32.17

© NSR 01 Aug 2008 - 31 Jul 2009

S : MECHANICAL SERVICES - PIPE SUPPLY SYSTEMS : WORKS OF ALTERATION/SMALL WORKS/REPAIR			Mat. £	Lab. £	Plant £	Total £
S2756						
602	20 mm Diameter	nr	26.32	11.71		38.03
604	25 mm Diameter	nr	36.46	12.38		48.84
606	32 mm Diameter	nr	61.78	13.05		74.83
608	40 mm Diameter	nr	76.86	14.73		91.59
610	50 mm Diameter	nr	117.88	16.73		134.61
612	65 mm Diameter	nr	240.62	18.19		258.81
614	80 mm Diameter	nr	340.23	21.31		361.54
616	100 mm Diameter	nr	366.65	23.39		390.04
S2758	**VALVES AND ANCILLARIES : supply and fit strainer**					
600	15 mm Diameter	nr	35.66	10.04		45.70
602	20 mm Diameter	nr	45.30	11.71		57.01
604	25 mm Diameter	nr	60.09	12.38		72.47
606	32 mm Diameter	nr	97.05	10.39		107.44
608	40 mm Diameter	nr	125.53	14.73		140.26
610	50 mm Diameter	nr	210.26	16.73		226.99
612	65 mm Diameter	nr	223.87	18.19		242.06
614	80 mm Diameter	nr	265.00	21.31		286.31
616	100 mm Diameter	nr	387.72	23.39		411.11
S2760	**VALVES AND ANCILLARIES : remove damaged draincock, supply and fix new**					
680	15 mm Diameter	nr	20.36	16.40		36.76
682	20 mm Diameter	nr	31.22	19.08		50.30
S2762	**VALVES AND ANCILLARIES : supply and fit drain cock**					
550	15 mm Hose outlet	nr	20.36	10.04		30.40
552	20 mm Hose outlet	nr	31.22	11.71		42.93
554	25 mm Hose outlet	nr	91.53	12.38		103.91
560	15 mm Hose union	nr	44.96	10.04		55.00
562	20 mm Hose union	nr	60.91	11.71		72.62
564	25 mm Hose union	nr	91.31	12.38		103.69
S2764	**VALVES AND ANCILLARIES : supply and fit air vent**					
590	8 mm Manual air vents	nr	1.65	4.02		5.67

© NSR 01 Aug 2008 - 31 Jul 2009

S29

S : MECHANICAL SERVICES - PIPE SUPPLY SYSTEMS : WORKS OF ALTERATION/SMALL WORKS/REPAIR			Mat. £	Lab. £	Plant £	Total £
S2764						
630	50 mm Diameter air bottle assemblies including drain pipe assembly	nr	149.26	35.14		184.40
635	15 mm Diameter automatic air vents including drain pipe assembly	nr	115.79	4.02		119.81
640	12 mm Diameter test plugs	nr	12.97	15.06		28.03
S2766	VALVES AND ANCILLARIES : air cock and vent					
600	Remove damaged 12 mm diameter air-cock, supply and fix new	nr	5.24	15.06		20.30
602	Remove damaged 12 mm diameter automatic air vent, supply and fix new	nr	117.04	15.06		132.10
S2768	VALVES AND ANCILLARIES : remove damaged test plug, supply and fix new					
670	12 mm Diameter	nr	12.97	22.76		35.73
S2770	VALVES AND ANCILLARIES : tighten or re-pack valve gland, grease spindle					
610	15 - 25 mm Diameter	nr		10.04		10.04
612	32 - 50 mm Diameter	nr		16.73		16.73
S2772	VALVES AND ANCILLARIES : supply and fit gauge					
650	100 mm Diameter dial pressure gauges, syphons and cocks	nr	91.85	15.06		106.91
652	150 mm Diameter dial pressure gauges, syphons and cocks	nr	108.98	15.06		124.04
654	100 mm Diameter dial thermometers including pockets	nr	165.50	15.06		180.56
656	150 mm Diameter dial thermometers including pockets	nr	186.25	15.06		201.31
658	100 mm Long straight stem thermometers including pockets	nr	112.18	15.06		127.24
660	100 mm Long angle stem thermometers including pockets	nr	112.20	15.06		127.26
	S60 : THERMAL INSULATION					
S6000	PIPEWORK : remove damaged hot water/heating service insulation to pipework and fittings, supply and fix new, to mild steel or copper pipework, concealed areas. Cotton scrim finish, rigid section, fixing with aluminium bands					
050	15 mm Diameter pipes	m	19.90	6.60		26.50
052	20 mm Diameter pipes	m	21.82	6.60		28.42
054	25 mm Diameter pipes	m	24.44	6.60		31.04
056	32 mm Diameter pipes	m	26.71	7.40		34.11
058	40 mm Diameter pipes	m	28.64	7.93		36.57
060	50 mm Diameter pipes	m	33.00	7.93		40.93

© NSR 01 Aug 2008 - 31 Jul 2009   S30

S : MECHANICAL SERVICES - PIPE SUPPLY SYSTEMS : WORKS OF ALTERATION/SMALL WORKS/REPAIR			Mat. £	Lab. £	Plant £	Total £
S6000						
062	65 mm Diameter pipes	m	38.14	10.57		48.71
064	80 mm Diameter pipes	m	42.22	11.89		54.11
066	100 mm Diameter pipes	m	56.11	13.21		69.32
S6010	PIPEWORK : supply and fit hot water/heating service insulation to pipework and fittings, mild steel or copper, concealed areas. Cotton scrim finish, rigid section, fixing with aluminium bands					
051	15 mm Diameter pipes	m	16.76	3.96		20.72
053	20 mm Diameter pipes	m	18.34	3.96		22.30
055	25 mm Diameter pipes	m	20.50	3.96		24.46
057	32 mm Diameter pipes	m	22.45	4.76		27.21
059	40 mm Diameter pipes	m	24.05	5.28		29.33
061	50 mm Diameter pipes	m	27.70	5.28		32.98
063	75 mm Diameter pipes	m	35.40	7.93		43.33
067	100 mm Diameter pipes	m	47.01	10.57		57.58
S6020	PIPEWORK : remove damaged cold water/chilled water service insulation to pipework fittings, supply and fix new to mild steel or copper pipework, concealed areas. Cotton scrim finish, rigid section, fixing with aluminium bands					
050	15 mm Diameter pipes	m	16.76	6.60		23.36
052	20 mm Diameter pipes	m	18.18	6.60		24.78
054	25 mm Diameter pipes	m	20.34	6.60		26.94
056	32 mm Diameter pipes	m	22.26	7.40		29.66
058	40 mm Diameter pipes	m	23.88	7.93		31.81
060	50 mm Diameter pipes	m	27.49	7.93		35.42
062	75 mm Diameter pipes	m	35.21	10.57		45.78
066	100 mm Diameter pipes	m	46.76	13.21		59.97
S6030	PIPEWORK : supply and fit cold water/chilled water service insulation to pipework and fittings, mild steel or copper, concealed areas. Cotton scrim finish, rigid section, fixing with aluminium bands					
051	15 mm Diameter pipes	m	16.60	3.96		20.56
053	20 mm Diameter pipes	m	18.18	3.96		22.14
055	25 mm Diameter pipes	m	20.34	3.96		24.30
057	32 mm Diameter pipes	m	22.26	4.76		27.02
059	40 mm Diameter pipes	m	23.88	5.28		29.16

© NSR 01 Aug 2008 - 31 Jul 2009

S31

S : MECHANICAL SERVICES - PIPE SUPPLY SYSTEMS : WORKS OF ALTERATION/SMALL WORKS/REPAIR			Mat. £	Lab. £	Plant £	Total £
S6030						
061	50 mm Diameter pipes	m	27.49	5.28		32.77
063	75 mm Diameter pipes	m	35.21	7.93		43.14
067	100 mm Diameter pipes	m	46.76	10.57		57.33
S6040	**PIPEWORK : remove damaged hot water/heating service insulation to pipework and fittings, supply and fix new to mild steel or copper pipework, exposed areas. Foil faced finish, rigid section, fixing with aluminium bands**					
050	15 mm Diameter pipes	m	16.98	6.60		23.58
052	20 mm Diameter pipes	m	21.53	6.60		28.13
054	25 mm Diameter pipes	m	22.76	6.60		29.36
056	32 mm Diameter pipes	m	24.11	7.40		31.51
058	40 mm Diameter pipes	m	26.36	7.93		34.29
060	50 mm Diameter pipes	m	27.89	7.93		35.82
062	65 mm Diameter pipes	m	32.01	10.57		42.58
064	80 mm Diameter pipes	m	39.56	11.89		51.45
066	100 mm Diameter pipes	m	51.55	13.21		64.76
S6050	**PIPEWORK : supply and fit hot water/heating service insulation to pipework and fittings, mild steel or copper, exposed areas. Foil faced finish, fixing with aluminium bands**					
051	15 mm Diameter pipes	m	21.53	3.96		25.49
053	20 mm Diameter pipes	m	21.53	3.96		25.49
055	25 mm Diameter pipes	m	22.76	3.96		26.72
057	32 mm Diameter pipes	m	24.11	4.76		28.87
059	40 mm Diameter pipes	m	26.36	5.28		31.64
061	50 mm Diameter pipes	m	27.89	5.28		33.17
063	65 mm Diameter pipes	m	32.01	7.93		39.94
065	80 mm Diameter pipes	m	39.56	9.25		48.81
067	100 mm Diameter pipes	m	51.55	10.57		62.12
S6060	**PIPEWORK : remove damaged cold water/chilled water service insulation to pipework fittings, supply and fix new, to mild steel or copper pipework, exposed areas. Foil faced finish, rigid section, fixing with aluminium bands**					
050	15 mm Diameter pipes	m	16.98	6.60		23.58
052	20 mm Diameter pipes	m	21.53	6.60		28.13
054	25 mm Diameter pipes	m	22.76	6.60		29.36

© NSR 01 Aug 2008 - 31 Jul 2009   S32

S : MECHANICAL SERVICES - PIPE SUPPLY SYSTEMS : WORKS OF ALTERATION/SMALL WORKS/REPAIR			Mat. £	Lab. £	Plant £	Total £
S6060						
056	32 mm Diameter pipes	m	24.11	7.40		31.51
058	40 mm Diameter pipes	m	26.36	7.93		34.29
060	50 mm Diameter pipes	m	27.89	7.93		35.82
062	65 mm Diameter pipes	m	32.01	10.57		42.58
064	80 mm Diameter pipes	m	39.56	11.89		51.45
066	100 mm Diameter pipes	m	51.55	13.21		64.76
S6070	**PIPEWORK : supply and fit cold water/chilled water service insulation to pipework and fittings, mild steel or copper, exposed areas. Foil faced finish, fixing with aluminium bands**					
051	15 mm Diameter pipes	m	21.26	3.96		25.22
053	20 mm Diameter pipes	m	21.26	3.96		25.22
055	25 mm Diameter pipes	m	22.76	3.96		26.72
057	32 mm Diameter pipes	m	24.11	4.76		28.87
059	40 mm Diameter pipes	m	26.36	5.28		31.64
061	50 mm Diameter pipes	m	27.89	5.28		33.17
063	65 mm Diameter pipes	m	32.01	7.93		39.94
065	80 mm Diameter pipes	m	39.56	9.25		48.81
067	100 mm Diameter pipes	m	51.55	10.57		62.12
S6080	**PIPEWORK : freeze pipework for isolation of services**					
100	15 mm Diameter pipes	nr	15.06	7.93		22.99
102	20 mm Diameter pipes	nr	15.06	10.57		25.63
104	25 mm Diameter pipes	nr	15.06	13.21		28.27
106	32 mm Diameter pipes	nr	30.12	17.17		47.29
108	40 mm Diameter pipes	nr	30.12	19.81		49.93
S6090	**HWS CYLINDERS : remove damaged insulation jacket to 45 litre domestic HWS cylinder, supply & fix new**					
900	Glass fibre filled, PVC covered insulating jacket, in strips, fixing with top wire ring and metal bands	nr	14.82	10.57		25.39
	S62 : FIRE PROTECTION SERVICES					
S6200	**HOSE REELS**					
010	Reposition roller hose guide including redrilling and plugging wall as necessary	nr		10.04		10.04

© NSR 01 Aug 2008 - 31 Jul 2009   S33

	S : MECHANICAL SERVICES - PIPE SUPPLY SYSTEMS : WORKS OF ALTERATION/SMALL WORKS/REPAIR		Mat. £	Lab. £	Plant £	Total £
S6200						
015	Remove damaged hose reel outlet box, fix only new box	nr		20.08		20.08
020	Supply and fix new outlet box	nr	23.63	15.06		38.69
S15	Remove damaged hose reel outlet box, Supply and fix new box	nr	23.63	20.08		43.71
S6202	HOSE REELS : remove damaged hose reel hose, fix only new hose					
030	24 Metre hose	nr		20.08		20.08
032	30 Metre hose	nr		23.43		23.43
034	36 Metre hose	nr		26.77		26.77
S6204	HOSE REELS : remove damaged hose reel, fix only new hose reel, wall or floor mounted					
036	24 Metre hose	nr		80.32		80.32
038	36 Metre hose	nr		90.36		90.36
S6206	HOSE REELS : fix only hose reels, non-automatic wall/floor mounted					
010	24 Metre hose	nr		61.91		61.91
012	36 Metre hose	nr		68.61		68.61
S6208	HOSE REELS : remove damaged hose reel, fix only new hose reel, recessed cabinet mounted					
036	24 Metre hose	nr		120.48		120.48
038	36 Metre hose	nr		128.85		128.85
S6210	HOSE REELS : fix only hose reels, automatic recessed door mounted					
010	24 Metre hose	nr		68.61		68.61
012	36 Metre hose	nr		76.97		76.97
S6212	HOSE REELS : supply only new hose reel hose					
030	24 Metre hose	nr	168.55			168.55
032	30 Metre hose	nr	168.55			168.55
034	36 Metre hose	nr	196.55			196.55
S6214	HOSE REELS : remove damaged hose reel hose, wall or floor mounted, supply and fix new					
S30	24 Metre hose	nr	168.55	20.08		188.63

© NSR 01 Aug 2008 - 31 Jul 2009

S : MECHANICAL SERVICES - PIPE SUPPLY SYSTEMS : WORKS OF ALTERATION/SMALL WORKS/REPAIR			Mat. £	Lab. £	Plant £	Total £
S6214						
S32	30 Metre hose,	nr	168.55	23.43		191.98
S34	36 Metre hose,	nr	196.55	26.77		223.32
S6216	HOSE REELS : supply only new hose reel, wall or floor mounted					
030	24 Metre hose	nr	507.58			507.58
032	30 Metre hose	nr	507.58			507.58
034	36 Metre hose	nr	507.58			507.58
S6218	HOSE REELS : remove damaged hose reel, wall or floor mounted, supply and fix new					
030	24 metre hose reel,	nr	168.55	80.32		248.87
032	30 metre hose reel,	nr	168.55	80.32		248.87
034	36 metre hose reel	nr	196.55	80.32		276.87
S6220	HOSE REELS : supply only new hose reel, recessed cabinet mounted					
030	24 Metre hose	nr	490.53			490.53
032	30 Metre hose	nr	490.53			490.53
034	36 Metre hose	nr	490.53			490.53
S6222	HOSE REELS : remove damaged hose reel, supply and fix new, recessed cabinet mounted					
030	24 Metre hose	nr	490.53	120.48		611.01
032	30 Metre hose reel,	nr	490.53	120.48		611.01
034	36 Metre hose reel,	nr	490.53	120.48		611.01
S6224	HOSE REELS : supply and fix hose reels, non-automatic wall/floor mounted					
010	24 metre hose	nr	507.58	31.18		538.76
012	36 metre hose	nr	507.58	31.18		538.76
S6226	HOSE REELS : supply and fix hose reels, automatic recessed door mounted					
010	24 metre hose	nr	490.53	46.77		537.30
012	36 metre hose	nr	490.53	46.77		537.30
S6228	DRY RISERS : fix only landing valve, flanged					
030	65 mm Diameter	nr		40.16		40.16

© NSR 01 Aug 2008 - 31 Jul 2009   S35

S : MECHANICAL SERVICES - PIPE SUPPLY SYSTEMS : WORKS OF ALTERATION/SMALL WORKS/REPAIR			Mat. £	Lab. £	Plant £	Total £
S6230	**DRY RISERS : supply only landing valve, flanged**					
030	65 mm Diameter	nr	315.57			315.57
S6232	**DRY RISERS : supply and fix landing valve, flanged**					
030	65 mm Diameter	nr	315.57	40.16		355.73
S6234	**DRY RISER : remove dry riser inlet breeching valve and final connections, set aside for re-use, fix only new box , refix valve and final connections**					
140	Two way	nr	371.66	100.40		472.06
142	Four way	nr	371.66	120.48		492.14
S6236	**DRY RISERS : fix only inlet breeching valve, flanged**					
035	Two way inlet	nr		60.24		60.24
S6238	**DRY RISERS : supply only inlet breeching valve, flanged**					
035	Two way inlet	nr	389.20			389.20
S6240	**DRY RISERS : supply and fix breeching valve, flanged**					
035	Two way inlet	nr	389.20	60.24		449.44
S6242	**DRY RISER : remove damaged dry riser inlet breeching valve, fix only new valve**					
140	Two way	nr		80.32		80.32
142	Four way	nr		204.95		204.95
S6244	**DRY RISER : remove damaged dry riser inlet breeching drain valve, fix only new valve**					
150	25 mm Diameter	nr		17.74		17.74
S6246	**DRY RISER**					
180	Remove damaged dry riser outlet box glass, fix only new glass	nr		13.39		13.39
182	Supply and fix new dry riser outlet box glass	nr	11.46	20.08		31.54
184	Remove damaged dry riser inlet box glass, fix only new glass, two way	nr		18.41		18.41
186	Remove damaged dry riser inlet box glass, fix only new glass, four way	nr		21.75		21.75
194	Remove dry riser landing valve and final connections, set aside for re-use, remove damaged outlet box, fix only new box, refix landing valve and final connections	nr	371.66	80.32		451.98
196	Supply and fix new outlet box	nr	371.66	20.08		391.74

© NSR 01 Aug 2008 - 31 Jul 2009   S36

S : MECHANICAL SERVICES - PIPE SUPPLY SYSTEMS : WORKS OF ALTERATION/SMALL WORKS/REPAIR			Mat. £	Lab. £	Plant £	Total £
S6246						
198	Remove damaged dry riser landing valve, fix only new valve 65 mm diameter	nr		46.18		46.18
S6248	**DRY RISER : supply only new glass for dry riser inlet box**					
140	Two way	nr	22.92			22.92
142	Four way	nr	45.84			45.84
S6250	**DRY RISER : remove damaged dry riser inlet box glass, supply and fix new**					
140	Two way,	nr	22.92	18.41		41.33
142	Four way	nr	45.84	21.75		67.59
S6252	**DRY RISER : supply only new dry riser inlet box**					
140	Two way	nr	371.66			371.66
142	Four way	nr	371.66			371.66
S6254	**DRY RISER : remove inlet breeching valve and final connections, set aside for re-use, supply and fix new box**					
140	Two way dry riser inlet breeching valve and final connections, set aside for re-use, supply and fix only new box , refix valve and final connections	nr	371.66	100.40		472.06
142	Four way	nr	371.66	120.48		492.14
S6256	**DRY RISER : supply only new dry riser valve**					
130	Supply only new dry riser inlet breeching drain valve 25mm diameter	nr	43.15			43.15
135	Supply only new dry riser landing valve 65 mm diameter	nr	315.57			315.57
170	Two way breeching valve	nr	389.20			389.20
172	Four way breeching valve	nr	778.40			778.40
S6258	**DRY RISER : remove damaged dry riser valve, supply and fit new**					
130	Inlet breeching drain valve, 25 mm diameter,	nr	43.15	17.74		60.89
170	Two way dry riser inlet breeching valve	nr	389.20	80.32		469.52
172	Four way dry riser inlet breeching valve,	nr	778.40	204.95		983.35
S6260	**PIPEWORK : repair leaking pipe or fitting**					
200	15 - 22 mm Diameter, copper	nr		9.37		9.37
202	28 - 35 mm Diameter, copper	nr		14.39		14.39

© NSR 01 Aug 2008 - 31 Jul 2009   S37

S : MECHANICAL SERVICES - PIPE SUPPLY SYSTEMS : WORKS OF ALTERATION/SMALL WORKS/REPAIR				Mat. £	Lab. £	Plant £	Total £
S6260							
204	42 - 54 mm Diameter, copper		nr		20.08		20.08
206	15 - 20 mm Diameter, galvanised mild steel		nr		11.04		11.04
208	25 - 32 mm Diameter, galvanised mild steel		nr		17.07		17.07
210	40 - 50 mm Diameter, galvanised mild steel		nr		24.10		24.10
S6262	PIPEWORK : supply and fit light gauge copper pipes (TX), EN1057 R250, capillary fittings, BS864 : Part 2						
010	15 mm Diameter		m	6.72	6.69		13.41
012		extra over for fittings with one end	nr	2.08	2.01		4.09
014		extra over for fittings with two ends	nr	0.60	3.01		3.61
016		extra over for fittings with three ends	nr	1.13	4.69		5.82
020	22 mm Diameter		m	13.44	8.03		21.47
022		extra over for fittings with one end	nr	3.88	2.68		6.56
024		extra over for fittings with two ends	nr	1.56	3.68		5.24
026		extra over for fittings with three ends	nr	3.61	6.02		9.63
030	28 mm Diameter		m	16.95	8.70		25.65
032		extra over for fittings with one end	nr	6.92	3.01		9.93
034		extra over for fittings with two ends	nr	3.08	4.35		7.43
036		extra over for fittings with three ends	nr	8.55	6.69		15.24
040	35 mm Diameter		m	39.37	10.04		49.41
042		extra over for fittings with one end	nr	15.30	4.02		19.32
044		extra over for fittings with two ends	nr	13.39	5.35		18.74
046		extra over for fittings with three ends	nr	21.80	9.04		30.84
050	42 mm Diameter		m	47.88	11.71		59.59
052		extra over for fittings with one end	nr	26.34	4.35		30.69
054		extra over for fittings with two ends	nr	22.13	6.36		28.49
056		extra over for fittings with three ends	nr	34.98	9.37		44.35
060	54 mm Diameter		m	61.60	13.39		74.99
062		extra over for fittings with one end	nr	36.77	4.69		41.46
064		extra over for fittings with two ends	nr	45.71	7.03		52.74
066		extra over for fittings with three ends	nr	70.53	10.04		80.57

© NSR 01 Aug 2008 - 31 Jul 2009   S38

S : MECHANICAL SERVICES - PIPE SUPPLY SYSTEMS : WORKS OF ALTERATION/SMALL WORKS/REPAIR

			Mat. £	Lab. £	Plant £	Total £
S6264	PIPEWORK : remove damaged pipework and fittings, supply and fix new, not exceeding 1000 mm long					
220	15 - 22 mm Diameter, copper	nr	19.74	27.21		46.95
222	28 - 35 mm Diameter, copper	nr	90.89	39.36		130.25
224	42 - 54 mm Diameter, copper	nr	164.04	54.41		218.45
226	15 - 20 mm Diameter, galvanised mild steel	nr	25.15	28.94		54.09
228	25 - 32 mm Diameter, galvanised mild steel	nr	49.06	36.47		85.53
230	40 - 50 mm Diameter, galvanised mild steel	nr	82.08	54.99		137.07
S6266	PIPEWORK : supply and fit light gauge copper pipes (TX), EN1057 R250, compression fittings, BS864 : Part 2					
010	15 mm Diameter	m	6.72	6.69		13.41
012	extra over for 15mm fittings with one end	nr	2.89	2.01		4.90
014	extra over for 15mm fittings with two ends	nr	2.23	3.01		5.24
016	extra over for 15mm fittings with three ends	nr	3.23	4.69		7.92
020	22 mm Diameter	m	13.44	8.03		21.47
022	extra over for 22mm fittings with one end	nr	3.46	2.68		6.14
024	extra over for 22mm fittings with two ends	nr	3.76	3.35		7.11
026	extra over for 22mm fittings with three ends	nr	5.28	5.02		10.30
030	28 mm Diameter	m	16.95	8.70		25.65
032	extra over for 28mm fittings with one end	nr	11.23	3.01		14.24
034	extra over for 28mm fittings with two ends	nr	13.49	3.68		17.17
036	extra over for 28mm fittings with three ends	nr	24.55	5.69		30.24
040	35 mm Diameter	m	39.37	10.04		49.41
042	extra over for 35mm fittings with one end	nr	20.27	3.68		23.95
044	extra over for 35mm fittings with two ends	nr	33.94	4.35		38.29
046	extra over for 35mm fittings with three ends	nr	44.87	6.69		51.56
050	42 mm Diameter	m	47.88	11.71		59.59
052	extra over for 42mm fittings with one end	nr	33.14	4.02		37.16
054	extra over for 42mm fittings with two ends	nr	47.51	4.69		52.20
056	extra over for 42mm fittings with three ends	nr	69.05	7.36		76.41
060	54 mm Diameter	m	61.60	13.39		74.99
062	extra over for 54mm fittings with one end	nr	46.22	5.02		51.24
064	extra over for 54mm fittings with two ends	nr	80.70	5.69		86.39
066	extra over for 54mm fittings with three ends	nr	111.12	8.37		119.49

© NSR 01 Aug 2008 - 31 Jul 2009

S39

S : MECHANICAL SERVICES - PIPE SUPPLY SYSTEMS : WORKS OF ALTERATION/SMALL WORKS/REPAIR				Mat. £	Lab. £	Plant £	Total £
S6268	PIPEWORK : supply and fit galvanised mild steel pipes, BS1387 : 1985, screwed fittings, BS143						
010	15 mm Diameter		m	5.22	8.37		13.59
012	extra over for fittings with one end		nr	10.02	5.02		15.04
014	extra over for fittings with two ends		nr	3.49	10.04		13.53
016	extra over for fittings with three ends		nr	28.63	14.06		42.69
020	20 mm Diameter		m	4.77	10.04		14.81
022	extra over for fittings with one end		nr	13.28	7.36		20.64
024	extra over for fittings with two ends		nr	4.03	13.39		17.42
026	extra over for fittings with three ends		nr	35.21	18.74		53.95
030	25 mm Diameter		m	7.61	11.71		19.32
032	extra over for fittings with one end		nr	21.32	8.37		29.69
034	extra over for fittings with two ends		nr	5.19	15.06		20.25
036	extra over for fittings with three ends		nr	56.74	21.08		77.82
040	32 mm Diameter		m	9.42	13.39		22.81
042	extra over for fittings with one end		nr	36.26	10.71		46.97
044	extra over for fittings with two ends		nr	7.63	17.74		25.37
046	extra over for fittings with three ends		nr	103.38	25.10		128.48
050	40 mm Diameter		m	10.95	15.06		26.01
052	extra over for fittings with one end		nr	36.26	10.71		46.97
054	extra over for fittings with two ends		nr	9.77	20.08		29.85
056	extra over for fittings with three ends		nr	103.38	28.11		131.49
060	50 mm Diameter		m	15.36	18.41		33.77
062	extra over for fittings with one end		nr	50.07	12.72		62.79
064	extra over for fittings with two ends		nr	14.65	24.10		38.75
066	extra over for fittings with three ends		nr	151.24	33.47		184.71
070	65 mm Diameter		m	20.84	19.08		39.92
072	Extra over for Cap 16W		nr	83.41	15.39		98.80
074	Extra over for Socket 14W		nr	25.09	29.45		54.54
076	Extra over for Female Tee 12W		nr	532.22	41.83		574.05
080	80 mm Diameter		m	97.15	24.43		121.58
082	Extra over for Cap 16W		nr	156.00	18.07		174.07
084	Extra over for Socket 14W		nr	35.08	34.81		69.89
086	Extra over for Female Tee 12W		nr	577.46	48.86		626.32
090	100 mm Diameter		m	38.16	33.13		71.29
© NSR 01 Aug 2008 - 31 Jul 2009							S40

	S : MECHANICAL SERVICES - PIPE SUPPLY SYSTEMS : WORKS OF ALTERATION/SMALL WORKS/REPAIR		Mat. £	Lab. £	Plant £	Total £
S6268						
092	Extra over for Cap 16W	nr	234.63	24.10		258.73
094	Extra over for Socket 14W	nr	66.98	48.19		115.17
096	Extra over for Female Tee 12W	nr	929.47	67.60		997.07
S6270	PIPEWORK : remove damaged hose reel final connections and set aside existing valves for re-use, supply and fix new pipework and fittings including setting as necessary, refix existing valves, not exceeding 1000 mm long					
240	22 mm Diameter, copper	nr	13.44	52.10		65.54
242	28 mm Diameter, copper	nr	16.95	57.88		74.83
244	20 mm Diameter, galvanised mild steel	nr	5.45	65.35		70.80
246	25 mm Diameter, galvanised mild steel	nr	7.61	72.35		79.96
S6272	PIPEWORK : supply and fit short water connections to hose reel, including pipework, fittings, isolating valve, setting as necessary, jointing to distribution pipework, not exceeding 1000 mm long					
108	22 mm Diameter copper	nr	31.49	21.42		52.91
110	28 mm Diameter copper	nr	37.40	24.10		61.50
112	20 mm Diameter mild steel	nr	18.48	28.11		46.59
114	25 mm Diameter mild steel	nr	25.66	25.10		50.76
S6274	PIPEWORK : supply and fit short water connections to landing valve, including pipework and fittings, jointing to distribution pipework, not exceeding 500 mm long					
116	65 mm Diameter mild steel	nr	106.41	110.44		216.85
S6276	ANCILLARIES : remove damaged hose reel hand wheel or lockshield valve, supply and fix new					
250	20 mm Diameter	nr	28.45	23.43		51.88
252	25 mm Diameter	nr	43.15	26.10		69.25
S6278	VALVES AND ANCILLARIES : hose reel service					
115	20 mm Isolating valves wheel-head and lockshield	nr	36.51	8.70		45.21
117	25 mm Isolating valves wheel-head and lockshield	nr	47.25	10.71		57.96
119	32 mm Isolating valves wheel-head and lockshield	nr	67.41	12.38		79.79
121	40 mm Isolating valves wheel-head and lockshield	nr	92.27	14.73		107.00
123	50 mm Isolating valves wheel-head and lockshield	nr	131.85	16.73		148.58
125	15 mm Globe valves wheel-head and lockshield	nr	58.37	7.36		65.73
© NSR 01 Aug 2008 - 31 Jul 2009						S41

S : MECHANICAL SERVICES - PIPE SUPPLY SYSTEMS : WORKS OF ALTERATION/SMALL WORKS/REPAIR			Mat. £	Lab. £	Plant £	Total £
S6278						
127	20 mm Globe valves wheel-head and lockshield	nr	28.45	8.70		37.15
129	25 mm Globe valves wheel-head and lockshield	nr	43.15	10.71		53.86
131	32 mm Globe valves wheel-head and lockshield	nr	65.49	12.38		77.87
133	40 mm Globe valves wheel-head and lockshield	nr	85.63	14.73		100.36
135	50 mm Globe valves wheel-head and lockshield	nr	130.26	16.73		146.99
137	65 mm Globe valves wheel-head and lockshield	nr	690.26	23.09		713.35
139	80 mm Globe valves wheel-head and lockshield	nr	967.86	60.57		1028.43
141	100 mm Globe valves wheel-head and lockshield	nr	734.34	65.26		799.60
143	20 mm Non-return valves	nr	32.11	8.70		40.81
145	25 mm Non-return valves	nr	47.16	10.71		57.87
147	32 mm Non-return valves	nr	71.82	12.38		84.20
149	40 mm Non-return valves	nr	93.58	14.73		108.31
151	50 mm Non-return valves	nr	140.23	16.73		156.96
	S76 : LABORATORY FITTINGS					
S7600	LABORATORY FITTINGS : remove damaged laboratory tap, single and multiple type, fix only new tap					
100	Bench mounted pattern	nr		20.08		20.08
102	Wall mounted pattern	nr		15.06		15.06
104	Remote control mounted pattern	nr		21.75		21.75
S7604	LABORATORY FITTINGS : fix only laboratory fitting, single and multiple type, all services					
020	Bench mounted pattern	nr		16.73		16.73
022	Wall mounted pattern	nr		11.71		11.71
024	Remote control mounted pattern	nr		18.41		18.41
S7608	LABORATORY FITTINGS : supply only laboratory tap					
100	Bench mounted pattern	nr	48.68			48.68
102	Wall mounted pattern	nr	102.12			102.12
104	Remote control mounted pattern	nr	139.61			139.61
S7612	LABORATORY FITTINGS : supply and fix laboratory fitting, single and multiple type, all services					
020	Bench mounted pattern	nr	48.68	16.73		65.41

© NSR 01 Aug 2008 - 31 Jul 2009   S42

S : MECHANICAL SERVICES - PIPE SUPPLY SYSTEMS : WORKS OF ALTERATION/SMALL WORKS/REPAIR			Mat. £	Lab. £	Plant £	Total £
S7612						
022	Wall mounted pattern	nr	102.12	11.71		113.83
024	Remote control pattern	nr	139.61	18.41		158.02
S7616	**LABORATORY FITTINGS : remove damaged laboratory tap, single and multiple type, supply and fix new**					
100	Bench mounted pattern	nr	48.68	20.08		68.76
102	Wall mounted pattern	nr	102.12	15.06		117.18
104	Remote control mounted pattern	nr	139.61	21.75		161.36
S7620	**PIPEWORK : repair leaking pipe or fitting not exceeding 1000 mm long**					
150	20 - 32 mm Diameter, polyethylene	nr		11.71		11.71
152	63 mm Diameter, polyethylene	nr		16.73		16.73
154	90 mm Diameter, polyethylene	nr		26.77		26.77
156	15 - 20 mm Diameter, black mild steel	nr		11.04		11.04
158	25 - 32 mm Diameter, black mild steel	nr		17.07		17.07
160	40 - 50 mm Diameter, black mild steel	nr		24.10		24.10
162	65 - 80 mm Diameter, black mild steel	nr		33.47		33.47
164	100 mm Diameter, black mild steel	nr		40.16		40.16
S7624	**PIPEWORK : supply and fit MDPE pipe and fittings, fusion fittings, Gas Corporation Standard, BGC/PS/PL2**					
090	20 mm Diameter	m	16.01	20.08		36.09
092	25 mm Diameter	m	16.42	21.75		38.17
094	32 mm Diameter	m	17.56	23.43		40.99
100	65 mm Diameter	m	38.06	30.12		68.18
S7628	**PIPEWORK : remove damaged pipework and fittings, supply and fix new, not exceeding 1000 mm long**					
156	15 - 20 mm Diameter, black mild steel	nr	4.77	33.80		38.57
158	25 - 32 mm Diameter, black mild steel	nr	8.65	42.17		50.82
160	40 - 50 mm Diameter, black mild steel	nr	14.01	63.25		77.26
S7632	**PIPEWORK : remove damaged final connections serving laboratory bench fitting, supply and fix new pipework and fittings including setting as necessary, not exceeding 1000 mm long**					
165	8 - 15 mm Pipework	nr	13.34	21.75		35.09

© NSR 01 Aug 2008 - 31 Jul 2009

S43

S : MECHANICAL SERVICES - PIPE SUPPLY SYSTEMS : WORKS OF ALTERATION/SMALL WORKS/REPAIR			Mat. £	Lab. £	Plant £	Total £
S7636	PIPEWORK : supply and fit light gauge copper pipe (TX), EN1057 R250, capillary fittings, BS864 : Part 2					
112	15 mm Diameter	m	10.36	20.08		30.44
114	22 mm Diameter	m	19.74	23.43		43.17
116	28 mm Diameter	m	22.33	26.10		48.43
118	35 mm Diameter	m	45.71	29.79		75.50
120	42 mm Diameter	m	58.16	33.47		91.63
122	54 mm Diameter	m	211.66	36.81		248.47
S7640	PIPEWORK : supply and fit black mild steel pipe, BS1387 : 1985, screwed fittings, BS143					
088	15 mm Diameter	m	32.83	52.54		85.37
090	20 mm Diameter	m	38.85	66.26		105.11
092	25 mm Diameter	m	55.68	74.63		130.31
094	32 mm Diameter	m	69.68	86.01		155.69
096	40 mm Diameter	m	82.45	98.73		181.18
098	50 mm Diameter	m	117.32	116.13		233.45
100	65 mm Diameter	m	232.39	140.23		372.62
S7644	PIPEWORK : supply and fit black mild steel pipe, BS1387 : 1985, weldable fittings, BS1965					
088	15 mm Diameter	m	13.59	47.52		61.11
090	20 mm Diameter	m	13.97	56.89		70.86
092	25 mm Diameter	m	19.19	67.94		87.13
094	32 mm Diameter	m	23.23	81.99		105.22
096	40 mm Diameter	m	24.82	96.38		121.20
098	50 mm Diameter	m	34.17	126.17		160.34
100	65 mm Diameter	m	45.01	161.31		206.32
S7648	PIPEWORK : supply and fit short connections to appliance, including pipework, fittings, setting as necessary, jointing to distribution pipework, not exceeding 1000 mm long					
124	12 mm Diameter	nr	9.89	23.43		33.32
126	15 mm Diameter	nr	10.36	26.77		37.13
S7652	VALVES AND ANCILLARIES : strip ballvalve, clean and renew O-ring					
198	15 - 50 mm Diameter	nr	1.76	10.04		11.80
© NSR 01 Aug 2008 - 31 Jul 2009						S44

S : MECHANICAL SERVICES - PIPE SUPPLY SYSTEMS : WORKS OF ALTERATION/SMALL WORKS/REPAIR			Mat. £	Lab. £	Plant £	Total £
S7656	VALVES AND ANCILLARIES : remove damaged ball valve, supply and fix new					
200	15 mm Diameter	nr	17.13	20.08		37.21
202	20 mm Diameter	nr	24.21	23.43		47.64
204	25 mm Diameter	nr	27.64	26.10		53.74
206	32 mm Diameter	nr	47.79	30.12		77.91
208	40 mm Diameter	nr	80.23	34.14		114.37
210	50 mm Diameter	nr	147.48	36.14		183.62
S7660	VALVES AND ANCILLARIES : recharging plug valve with sealing compound					
220	15 - 50 mm Diameter	nr		18.07		18.07
222	65 - 100 mm Diameter	nr		30.12		30.12
S7664	VALVES AND ANCILLARIES : strip plug valve and clean, recharge sealing compound					
220	15 - 50 mm Diameter	nr		20.08		20.08
222	65 - 100 mm Diameter	nr		33.13		33.13
S7668	VALVES AND ANCILLARIES : remove damaged plug valve, supply and fix new					
200	15 mm Diameter	nr	90.19	20.08		110.27
202	20 mm Diameter	nr	107.80	23.43		131.23
204	25 mm Diameter	nr	131.49	26.10		157.59
208	40 mm Diameter	nr	221.20	34.14		255.34
210	50 mm Diameter	nr	299.94	36.14		336.08
212	65 mm Diameter	nr	299.94	46.18		346.12
214	80 mm Diameter	nr	467.64	54.22		521.86
216	100 mm Diameter	nr	662.87	60.24		723.11
S7672	VALVES AND ANCILLARIES : supply and fit plug valve, square head					
126	15 mm Diameter	nr	90.19	10.04		100.23
128	20 mm Diameter	nr	107.80	11.71		119.51
130	25 mm Diameter	nr	131.49	13.05		144.54
134	40 mm Diameter	nr	221.20	17.07		238.27
136	50 mm Diameter	nr	299.94	18.07		318.01
138	80 mm Diameter	nr	467.64	23.09		490.73

© NSR 01 Aug 2008 - 31 Jul 2009

S45

S : MECHANICAL SERVICES - PIPE SUPPLY SYSTEMS : WORKS OF ALTERATION/SMALL WORKS/REPAIR			Mat. £	Lab. £	Plant £	Total £
S7676	VALVES AND ANCILLARIES : supply and fit ball valve					
126	15 mm Diameter	nr	9.42	10.04		19.46
128	20 mm Diameter	nr	13.03	11.71		24.74
130	25 mm Diameter	nr	27.64	13.05		40.69
132	32 mm Diameter	nr	47.79	15.06		62.85
134	40 mm Diameter	nr	80.23	17.07		97.30
136	50 mm Diameter	nr	147.48	18.07		165.55
S7680	VALVES AND ANCILLARIES : supply and fit gas governor					
126	15 mm Diameter, fast open	nr	73.04	10.04		83.08
128	20 mm Diameter, fast open	nr	87.94	11.71		99.65
130	25 mm Diameter, fast open	nr	148.80	13.05		161.85
134	40 mm Diameter, slow open	nr	898.95	17.07		916.02
136	50 mm Diameter, slow open	nr	879.06	18.07		897.13
138	65 mm Diameter, slow open	nr	1151.02	23.09		1174.11
	S85 : WATER TREATMENT					
S8500	WATER SOFTENERS : supply and fix					
000	1.6 cu.m simplex water softener, timer control	nr	516.00	228.83		744.83
002	2.3 cu.m simplex water softener, timer control	nr	540.00	260.76		800.76
004	3.3 cu. m simplex water sotener, timer control	nr	588.00	274.07		862.07
006	5.0 cu.m simplex water softener, timer control	nr	659.00	303.34		962.34
008	7.5 cu.m simplex water softener, timer control	nr	735.00	313.98		1048.98
010	11.6 cu.m simplex water softener, timer control	nr	1047.00	335.27		1382.27
012	16.6 cu.m simplex water softener, timer control	nr	1578.00	369.86		1947.86
014	5.0 cu. m Duplex, volumetric control	nr	1536.00	332.60		1868.60
016	7.5 cu. m Duplex, volumetric control	nr	1734.00	340.59		2074.59
018	11.6 cu. m Duplex, volumetric control	nr	2271.00	377.84		2648.84
020	16.6 cu. m Duplex, volumetric control	nr	2840.00	409.77		3249.77
S8502	WATER SCALE INHIBITER : supply and fix in-line lime scale inhibitor					
000	15 mm	nr	48.80	9.04		57.84
002	22 mm	nr	52.22	11.04		63.26
004	28 mm	nr	300.00	13.05		313.05
006	35 mm	nr	360.00	15.06		375.06

© NSR 01 Aug 2008 - 31 Jul 2009   S46

S : MECHANICAL SERVICES - PIPE SUPPLY SYSTEMS : WORKS OF ALTERATION/SMALL WORKS/REPAIR			Mat. £	Lab. £	Plant £	Total £
S8502						
008	42 mm	nr	470.00	20.75		490.75
010	54 mm	nr	480.00	24.77		504.77
S8504	**WATER CONDITIONERS : supply and fix in line electro magnetic water conditioner**					
000	0.75 in (20 mm) diameter	nr	2790.00	383.16		3173.16
002	1.0 in (25 mm) diameter	nr	3150.00	399.13		3549.13
006	1.25 in (32 mm) diameter	nr	3971.12	409.77		4380.89
008	1.5 in (40 mm) diameter	nr	4270.00	491.91		4761.91
010	2.0 in (50 mm) diameter	nr	4840.00	447.02		5287.02
012	3.0 in (80 mm) diameter	nr	5130.00	462.99		5592.99
014	4.0 in (100 mm) diameter	nr	5410.00	510.88		5920.88
	S90 : SUMP AND SUBMERSIBLE PUMPS					
S9006	**SUMP PUMPS: supply and fix sump pump, with float switch and 5 metres of cable.**					
000	0.22 kW, single phase	nr	234.13	18.41		252.54
002	0.45 kW, single phase	nr	304.24	18.41		322.65
004	0.6 kW, single phase	nr	383.60	18.41		402.01
006	0.75 kW, single phase	nr	403.45	18.41		421.86
S9008	**SUMP PUMPS: supply and fix sump pump, no float switch fitted**					
000	1.5 hp, single phase, 1.25 in diameter connector	nr	875.00	57.88		932.88
002	1.5 hp, three phase, 1.25 in diameter connector	nr	888.00	63.67		951.67
004	2.0 hp, single phase, 1.25 in diameter connector	nr	866.42	57.88		924.30
006	2.0 hp, three phase, 1.25 in diameter connector	nr	839.96	63.67		903.63
008	2.5 hp, single phase, 1.25 in diameter connector	nr	787.05	57.88		844.93
010	2.5 hp, three phase, 1.25 in diameter connector	nr	501.00	63.67		564.67
012	3.0 hp, single phase, 1.25 in diameter connector	nr	1111.13	63.67		1174.80
014	3.0 hp, three phase, 1.25 in diameter connector	nr	1084.67	69.46		1154.13
S9010	**SUBMERSIBLE PUMPS: supply and fix submersible pump with 10 metres cable**					
000	0.25kW, single phase, 1.25 in diameter connection, float switch	nr	210.00	43.41		253.41
002	0.55 kW, single phase, 1.25 in diameter connection, float switch	nr	279.00	43.41		322.41

© NSR 01 Aug 2008 - 31 Jul 2009

S47

S : MECHANICAL SERVICES - PIPE SUPPLY SYSTEMS : WORKS OF ALTERATION/SMALL WORKS/REPAIR			Mat. £	Lab. £	Plant £	Total £
S9010						
004	0.75 kW, single phase, 1.5 in diameter connection, float switch	nr	443.00	43.41		486.41
006	1.1 kW, single phase, 2.0 in diameter connection, float switch	nr	558.00	49.20		607.20
008	0.55 kW, three phase, 1.25 in diameter connection	nr	279.00	49.20		328.20
010	0.55 kW, three phase, 1.5 in diameter connection	nr	414.00	49.20		463.20
012	0.75 kW, three phase, 1.5 in diameter connection	nr	443.00	49.20		492.20
014	1.1 kW, three phase, 1.5 in diameter connection	nr	530.00	53.83		583.83
016	1.5 kW, three phase, 1.5 in diameter connection	nr	578.56	53.83		632.39
018	0.75 kW, three phase, 2.0 in diameter connection	nr	455.00	59.62		514.62
020	1.1 kW, three phase, 2.0 in diameter connection	nr	558.00	59.62		617.62
022	1.5 kW, three phase, 2.0 in diameter connection	nr	621.55	59.62		681.17

© NSR 01 Aug 2008 - 31 Jul 2009

S48

			Mat. £	Lab. £	Plant £	Total £
	T : MECHANICAL SERVICES - MECHANICAL HEATING/COOLING REFRIGERATION SYSTEMS : WORKS OF ALTERATION/SMALL WORKS/REPAIR					
	T15 : HOT AND COLD WATER SERVICES : PLANT AND EQUIPMENT					
T1500	**WATER HEATERS**					
200	Remove damaged electric water heater thermal cut-out, fix only new cut-out	nr		10.37		10.37
202	Supply and fix new electric water heater thermal cut out	nr	10.35	7.03		17.38
204	Remove damaged electric water heater thermostat, fix only new thermostat	nr		14.06		14.06
206	Supply and fix new electric water heater thermostat	nr	8.25	10.37		18.62
208	Remove damaged electric water heater element, fix only new element	nr		14.06		14.06
210	Supply and fix new electric water heater element	nr	43.43	10.37		53.80
212	Remove damaged water heater valve, fix only new valve	nr		14.06		14.06
214	Supply and fix new water heater valve	nr	5.23	10.37		15.60
216	Remove electric water heater damaged swivel spout, fix only new spout	nr		14.06		14.06
218	Supply and fix new electric water heater swivel spout	nr	38.63	10.37		49.00
220	Refix electric water heater including redrilling and plugging wall as necessary	nr		60.24		60.24
T1512	**WATER HEATERS : remove damaged electric water heater, fix only new heater, over or under sink application**					
236	10 Litre capacity	nr		34.15		34.15
238	15 Litre capacity	nr		34.15		34.15
T1514	**WATER HEATERS : remove, supply and fix water heater, over or under sink application**					
236	10 Litre capacity	nr	276.64	34.15		310.79
238	15 Litre capacity	nr	351.52	34.15		385.67
T1518	**WATER HEATERS : supply only new heater, over or under sink application**					
236	10 Litre capacity	nr	276.64			276.64
238	15 Litre capacity	nr	351.52			351.52
T1524	**WATER HEATERS : fix only electric water heater, under or over sink type, swivel spout, thermostatically controlled**					
088	10 Litre capacity	nr		15.06		15.06

© NSR 01 Aug 2008 - 31 Jul 2009　　　　　　　　　　　　　　　　T1

	T : MECHANICAL SERVICES - MECHANICAL HEATING/COOLING REFRIGERATION SYSTEMS : WORKS OF ALTERATION/SMALL WORKS/REPAIR		Mat. £	Lab. £	Plant £	Total £
T1524						
090	15 Litre capacity	nr		15.06		15.06
T1530	WATER HEATERS : supply and fix electric water heater, under or over sink type, swivel spout, thermostatically controlled					
088	10 litre capacity	nr	276.64	15.06		291.70
090	15 litre capacity	nr	351.52	15.06		366.58
T1536	WATER HEATERS : remove damaged electric water heater, fix only new heater, electronic or immersion cistern type					
242	23 - 55 Litre capacity	nr		81.04		81.04
244	68 - 114 Litre capacity	nr		104.19		104.19
T1542	WATER HEATERS : supply only new electric water heater, electronic or immersion cistern type					
242	23 - 55 Litre capacity	nr	426.30			426.30
244	68 - 114 Litre capacity	nr	610.00			610.00
T1548	WATER HEATERS : remove damaged electric water heater, supply and fix new heater, electronic or immersion cistern type					
242	23 - 55 Litre capacity	nr	581.70	81.04		662.74
244	68 - 114 litre capacity	nr	610.00	104.19		714.19
T1554	WATER HEATERS : fix only electric water heater, cistern type, electronic or immersion					
094	23 Litre capacity	nr		10.04		10.04
096	34 Litre capacity	nr		26.05		26.05
098	55 Litre capacity	nr		26.05		26.05
100	68 Litre capacity	nr		34.73		34.73
102	91 Litre capacity	nr		34.73		34.73
104	114 Litre capacity	nr		40.52		40.52
T1560	WATER HEATERS : supply only electric water heater, electronic cistern type					
094	23 Litre capacity	nr	983.00			983.00
096	34 Litre capacity	nr	983.00			983.00
098	55 Litre capacity	nr	983.00			983.00
100	68 Litre capacity	nr	1123.45			1123.45
102	91 Litre capacity	nr	1123.45			1123.45

© NSR 01 Aug 2008 - 31 Jul 2009

T2

T : MECHANICAL SERVICES - MECHANICAL HEATING/COOLING REFRIGERATION SYSTEMS : WORKS OF ALTERATION/SMALL WORKS/REPAIR			Mat. £	Lab. £	Plant £	Total £
T1566	WATER HEATERS : supply and fix electric water heater, electronic cistern type					
094	23 litre capacity	nr	983.00	10.04		993.04
096	34 litre capacity	nr	983.00	26.05		1009.05
098	55 litre capacity	nr	983.00	26.05		1009.05
100	68 litre capacity	nr	1123.45	34.73		1158.18
102	91 litre capacity	nr	1123.45	34.73		1158.18
T1572	WATER HEATERS : supply only electric water heater, immersion cistern type					
094	30 Litre capacity	nr	355.95			355.95
097	50 litre capacity	nr	426.30			426.30
100	75 Litre capacity	nr	581.70			581.70
104	150 Litre capacity	nr	610.00			610.00
T1578	WATER HEATERS : supply and fix electric water heater, cistern type, immersion					
094	30 litre capacity	nr	355.95	10.04		365.99
097	50 litre capacity	nr	426.30	26.05		452.35
100	75 litre capacity	nr	581.70	34.73		616.43
104	150 litre capacity	nr	610.00	40.52		650.52
T1584	WATER HEATERS: fix only multipoint gas water heater					
150	Wall mounted heater, over sink type	nr		80.32		80.32
	T25 : SPACE HEATING SERVICES : PLANT AND EQUIPMENT					
T2510	AIR CURTAINS: Supply only 3 Speed electric remote control air curtain.					
100	6kW, 1000mm long	nr	736.04			736.04
110	9kW, 1500mm long	nr	913.79			913.79
120	12kW, 2000mm long	nr	1002.38			1002.38
T2511	AIR CURTAINS: Supply and fix 3 Speed electric remote control air curtain.					
100	6kW, 1000mm long	nr	736.04	40.16		776.20
110	9kW, 1500mm long	nr	913.79	45.18		958.97
120	12kW, 2000mm long	nr	1002.38	50.20		1052.58

© NSR 01 Aug 2008 - 31 Jul 2009 — T3

T : MECHANICAL SERVICES - MECHANICAL HEATING/COOLING REFRIGERATION SYSTEMS : WORKS OF ALTERATION/SMALL WORKS/REPAIR			Mat. £	Lab. £	Plant £	Total £
T2512	AIR CURTAINS: Fix only 3 Speed electric remote control air curtain.					
100	6kW, 1000mm long	nr		40.16		40.16
110	9kW, 1500mm long	nr		45.18		45.18
120	12kW, 2000mm long	nr		50.20		50.20
T2518	BOILERS					
008	Fix only boiler and flue, floor mounted, gas fired, conventional flue, not exceeding 8000 mm long, 0 - 25 kW range	nr		243.11		243.11
011	Fix only boiler and flue, wall mounted, gas fired, balanced flue, 0 - 25 kW range	nr		208.38		208.38
017	Fix only boiler and flue, floor mounted, oil fired, pressure jet burner, conventional flue, not exceeding 8000 mm long, 0 - 25 kW range	nr		272.05		272.05
020	Fix only boiler and flue, floor mounted, oil fired, vapouriser burner, balanced flue, 0 - 25 kW range	nr		208.38		208.38
023	Supply only boiler and flue, floor mounted, gas fired, balanced flue, 0 - 25 kW range	nr	1031.67			1031.67
026	Supply only boiler and flue, floor mounted, gas fired, conventional flue, not exceeding 8000 mm long, 0 - 25 kW range	nr	1031.67			1031.67
029	Supply only boiler and flue, wall mounted, gas fired, balanced flue, 0 - 25 kW range	nr	901.25			901.25
031	Supply only boiler and flue, wall mounted, gas fired, conventional flue, not exceeding 8000 mm long, 0 - 25 kW range	nr	1171.72			1171.72
034	Supply only boiler and flue, floor mounted, oil fired, conventional flue, not exceeding 8000 mm long, 0 - 25 kW range	nr	2043.68			2043.68
037	Supply only boiler and flue, floor mounted, oil fired, balanced flue, 0 - 25 kW range	nr	2162.34			2162.34
T2519	BOILERS					
023	Supply and fix boiler and flue, floor mounted, gas fired, balanced flue, 0 - 25 kW range	nr	761.20	208.38		969.58
026	Supply and fix boiler and flue, floor mounted, gas fired, conventional flue, not exceeding 8000 mm long, 0 - 25 kW range	nr	1031.67	272.05		1303.72
029	Supply and fix boiler and flue, wall mounted, gas fired, balanced flue, 0 - 25 kW range	nr	901.25	208.38		1109.63
031	Supply and fix boiler and flue, wall mounted, gas fired, conventional flue, not exceeding 8000 mm long, 0 - 25 kW range	nr	1171.72	243.11		1414.83
034	Supply and fix boiler and flue, floor mounted, oil fired, conventional flue, not exceeding 8000 mm long, 0 - 25 kW range	nr	2043.68	243.11		2286.79
037	Supply and fix boiler and flue, floor mounted, oil fired, balanced flue, 0 - 25 kW range	nr	2162.34	208.38		2370.72

© NSR 01 Aug 2008 - 31 Jul 2009

T : MECHANICAL SERVICES - MECHANICAL HEATING/COOLING REFRIGERATION SYSTEMS : WORKS OF ALTERATION/SMALL WORKS/REPAIR			Mat. £	Lab. £	Plant £	Total £
T2520	**CIRCULATING PUMPS**					
065	Fix only circulating pump, bronze, in line glandless, 20 - 50 mm diameter line size	nr		60.78		60.78
068	Fix only circulating pump, cast iron, in line, centrifugal, 32 - 50 mm diameter line size	nr		86.83		86.83
071	Fix only circulating pump, cast iron, in line, centrifugal twin set, 32 - 50 mm diameter line size	nr		138.92		138.92
074	Supply only circulating pump, bronze, in line glandless, 20 - 50 mm diameter line size	nr	169.48			169.48
077	Supply only circulating pump, cast iron, in line, centrifugal, 32 - 50 mm diameter line size	nr	409.00			409.00
080	Supply only circulating pump, cast iron, in line, centrifugal twin set, 32 - 50 mm diameter line size	nr	1183.00			1183.00
T2521	**CIRCULATING PUMPS**					
074	Supply and fix circulating pump, bronze, in line glandless, 20 - 50 mm diameter line size	nr	169.48	60.78		230.26
077	Supply and fix circulating pump, cast iron, in line, centrifugal, 32 - 50 mm diameter line size	nr	409.00	86.83		495.83
080	Supply and fix circulating pump, cast iron, in line, centrifugal twin set, 32 - 50 mm diameter line size	nr	1183.00	138.92		1321.92
T2522	**RADIATORS : remove steel panel radiator and set aside for re-use, remove support brackets, refix brackets and radiator**					
080	1000 mm Long	nr		34.73		34.73
082	1000 - 1500 mm Long	nr		37.62		37.62
084	1500 - 2000 mm Long	nr		40.52		40.52
086	2000 - 2500 mm Long	nr		52.10		52.10
T2523	**RADIATORS : remove steel panel radiator and set aside for re-use, remove damaged support brackets, fix only new brackets, refix radiator**					
080	1000 mm Long	nr		34.73		34.73
082	1000 - 1500 mm Long	nr		37.62		37.62
084	1500 - 2000 mm Long	nr		40.52		40.52
086	2000 - 2500 mm Long	nr		52.10		52.10
T2524	**RADIATORS : fix only radiator, steel panel, air vent, plug, brackets, average height 300 - 700 mm**					
083	1000 mm Long	nr		34.73		34.73
085	1000 - 1500 mm Long	nr		40.52		40.52
087	1500 - 2000 mm Long	nr		48.62		48.62

© NSR 01 Aug 2008 - 31 Jul 2009   T5

T : MECHANICAL SERVICES - MECHANICAL HEATING/COOLING REFRIGERATION SYSTEMS : WORKS OF ALTERATION/SMALL WORKS/REPAIR			Mat. £	Lab. £	Plant £	Total £
T2524						
088	2000 - 2500 mm Long	nr		53.83		53.83
T2525	RADIATORS : remove damaged / leaking steel panel radiator, remove and set aside for re-use valve tails, fix only new radiator, air cock and plug, refix valve tails, average height 300 - 700 mm					
080	1000 mm Long	nr		52.10		52.10
082	1000 - 1500 mm Long	nr		52.10		52.10
084	1500 - 2000 mm Long	nr		52.10		52.10
086	2000 - 2500 mm Long	nr		69.46		69.46
T2526	RADIATORS : supply and fix radiator, steel panel, air vent, plug, brackets, average height 300 - 700 mm					
090	400 mm long (K1/P1)	nr	36.59	34.73		71.32
092	500 mm long (K1/P1)	nr	45.73	34.73		80.46
093	600 mm long (K1/P1)	nr	54.88	34.73		89.61
095	700 mm long (K1/P1)	nr	64.01	34.73		98.74
096	800 mm long (K1/P1)	nr	73.16	34.73		107.89
098	900 mm long (K1/P1)	nr	94.20	34.73		128.93
100	1000 mm long (K1/P1)	nr	36.59	34.73		71.32
102	1000 - 1500 mm long (K1/P1)	nr	168.94	34.73		203.67
104	1500 - 2000 mm long (K1/P1)	nr	253.41	41.68		295.09
106	2000 - 2500 mm long (K1/P1)	nr	309.73	48.04		357.77
110	400 mm long (K2/P+)	nr	76.35	34.73		111.08
112	500 mm long (K2/P+)	nr	95.41	34.73		130.14
114	600 mm long (K2/P+)	nr	114.49	34.73		149.22
116	700 mm long (K2/P+)	nr	133.59	34.73		168.32
118	800 mm long (K2/P+)	nr	152.68	34.73		187.41
120	900 mm long (K2/P+)	nr	171.75	34.73		206.48
122	1000 - 1500 mm long (K2/P+)	nr	307.98	34.73		342.71
124	1500 - 2000 mm long (K2/P+)	nr	461.98	41.68		503.66
126	2000 - 2500 mm long (K2/P+)	nr	564.64	48.04		612.68
T2527	RADIATORS : refix existing steel panel radiator					
080	1000 mm Long	nr		17.37		17.37
082	1000 - 1500 mm Long	nr		20.26		20.26

© NSR 01 Aug 2008 - 31 Jul 2009   T6

T : MECHANICAL SERVICES - MECHANICAL HEATING/COOLING REFRIGERATION SYSTEMS : WORKS OF ALTERATION/SMALL WORKS/REPAIR			Mat. £	Lab. £	Plant £	Total £
T2527						
084	1500 - 2000 mm Long	nr		23.15		23.15
086	2000 - 2500 mm Long	nr		27.21		27.21
T2528	RADIATORS : supply only steel panel radiator complete with brackets, air-cock and plug, average height 300 - 700 mm					
090	400 mm Long (K1/P1)	nr	36.59			36.59
092	500 mm Long (K1/P1)	nr	45.73			45.73
093	600 mm Long (K1/P1)	nr	54.88			54.88
095	700 mm Long (K1/P1)	nr	64.01			64.01
096	800 mm Long (K1/P1)	nr	73.16			73.16
098	900 mm Long (K1/P1)	nr	94.20			94.20
100	1000 mm Long (K1/P1)	nr	104.68			104.68
102	1000 - 1500 mm Long (K1/P1)	nr	168.94			168.94
104	1500 - 2000 mm Long (K1/P1)	nr	253.41			253.41
106	2000 - 2500 mm Long (K1/P1)	nr	309.73			309.73
110	400 mm Long (K2/P+)	nr	76.35			76.35
112	500 mm Long (K2/P+)	nr	95.41			95.41
114	600 mm Long (K2/P+)	nr	114.49			114.49
116	700 mm Long (K2/P+)	nr	133.59			133.59
118	800 mm Long (K2/P+)	nr	152.68			152.68
120	900 mm Long (K2/P+)	nr	171.75			171.75
122	1000 - 1500 mm Long (K2/P+)	nr	307.98			307.98
124	1500 - 2000 mm Long (K2/P+)	nr	461.98			461.98
126	2000 - 2500 mm Long (K2/P+)	nr	564.64			564.64
T2529	RADIATORS : remove damaged / leaking steel panel radiator, remove and set aside for re-use valve tails, supply steel panel radiator complete with brackets, air-cock and plug, fix new radiator, air cock and plug, refix valve tails, average height 300 - 700 mm					
090	400 mm. long (K1 /P1)	nr	36.59	52.10		88.69
092	500 mm. long (K1 /P1)	nr	45.73	52.10		97.83
093	600 mm. long (K1 /P1)	nr	54.88	52.10		106.98
095	700 mm. long (K1 /P1)	nr	64.01	52.10		116.11
096	800 mm. long (K1 /P1)	nr	73.16	52.10		125.26
098	900 mm. long (K1 /P1)	nr	94.20	52.10		146.30

© NSR 01 Aug 2008 - 31 Jul 2009

T7

T : MECHANICAL SERVICES - MECHANICAL HEATING/COOLING REFRIGERATION SYSTEMS : WORKS OF ALTERATION/SMALL WORKS/REPAIR			Mat. £	Lab. £	Plant £	Total £
T2529						
100	1000 mm. long (K1/P1)	nr	104.68	52.10		156.78
102	1000 - 1500 mm. long (K1/P1)	nr	168.94	69.46		238.40
104	1500 - 2000 mm. long (K1/P1)	nr	253.41	86.83		340.24
106	2000 - 2500 mm. long (K1/P1)	nr	309.73	104.19		413.92
110	400 mm. long (K2 /P+)	nr	76.35	52.10		128.45
112	500 mm. long (K2 /P+)	nr	95.41	52.10		147.51
114	600 mm. long (K2 /P+) mm	nr	114.49	52.10		166.59
116	700 mm. long (K2 /P+)	nr	133.59	52.10		185.69
118	800 mm. long (K2 /P+)	nr	152.68	52.10		204.78
120	900 mm. long (K2 /P+)	nr	171.75	52.10		223.85
122	1000 - 1500 mm. long (K2 /P+)	nr	307.98	69.46		377.44
124	1500 - 2000 mm. long (K2 /P+)	nr	461.98	86.83		548.81
126	2000 - 2500 mm. long (K2 /P+)	nr	564.64	104.19		668.83
T2530	RADIATORS : remove damaged / leaking cast iron radiator, remove and set aside for re-use valve tails, fix only new radiator, air cock and plug, refix valve tails, average height 300 - 700 mm					
080	1000 mm Long	nr		69.46		69.46
082	1000 - 1500 mm Long	nr		69.46		69.46
084	1500 - 2000 mm Long	nr		69.46		69.46
086	2000 - 2500 mm Long	nr		86.83		86.83
T2531	RADIATORS : remove cast iron radiator and set aside for re-use, remove support brackets, refix brackets and radiator					
080	1000 mm Long	nr		86.83		86.83
082	1000 - 1500 mm Long	nr		98.40		98.40
084	1500 - 2000 mm Long	nr		105.93		105.93
086	2000 - 2500 mm Long	nr		117.50		117.50
T2533	RADIATORS : refix existing cast iron radiator					
080	1000 mm Long	nr		26.05		26.05
082	1000 - 1500 mm Long	nr		26.05		26.05
084	1500 - 2000 mm Long	nr		35.89		35.89
086	2000 - 2500 mm Long	nr		35.89		35.89

© NSR 01 Aug 2008 - 31 Jul 2009   T8

T : MECHANICAL SERVICES - MECHANICAL HEATING/COOLING REFRIGERATION SYSTEMS : WORKS OF ALTERATION/SMALL WORKS/REPAIR			Mat. £	Lab. £	Plant £	Total £
T2534	RADIATORS : fix only radiator, cast iron column, air vent, plug, brackets, average height 300 - 700 mm					
083	1000 mm Long	nr		52.10		52.10
085	1000 - 1500 mm Long	nr		52.10		52.10
087	1500 - 2000 mm Long	nr		61.94		61.94
088	2000 - 2500 mm Long	nr		71.78		71.78
T2535	RADIATORS : supply only cast-iron radiator complete with brackets, air-cock and plug, average height 300 - 700 mm					
091	524 mm. Long	nr	377.00			377.00
094	624 mm Long	nr	428.00			428.00
097	674 mm Long	nr	454.00			454.00
099	724 mm Long	nr	479.00			479.00
101	824 mm Long	nr	530.00			530.00
103	874 mm Long	nr	555.00			555.00
105	1024 mm Long	nr	632.00			632.00
T2536	RADIATORS : remove damaged/leaking cast iron radiator, remove and set aside for re-use valve tails, supply new cast-iron radiator complete with brackets, air cock and plug, fix new radiator, air cock and plug, refix valve					
091	524 mm long	nr	377.00	69.46		446.46
094	624 mm long	nr	428.00	69.46		497.46
097	674 mm long	nr	454.00	69.46		523.46
099	724 mm long	nr	479.00	69.46		548.46
101	824 mm long	nr	530.00	86.83		616.83
103	874 mm long	nr	555.00	86.83		641.83
105	1024 mm long	nr	632.00	86.83		718.83
T2537	RADIATORS : supply and fix radiator, cast iron column, air vent, plug, brackets, average height 300 - 700 mm					
091	524 mm long	nr	377.00	52.10		429.10
094	624 mm long	nr	428.00	52.10		480.10
097	674 mm long	nr	454.00	52.10		506.10
099	724 mm long	nr	479.00	52.10		531.10
101	824 mm long	nr	530.00	52.10		582.10
103	874 mm long	nr	555.00	61.94		616.94
105	1024 mm long	nr	632.00	71.78		703.78

© NSR 01 Aug 2008 - 31 Jul 2009

T9

T : MECHANICAL SERVICES - MECHANICAL HEATING/COOLING REFRIGERATION SYSTEMS : WORKS OF ALTERATION/SMALL WORKS/REPAIR			Mat. £	Lab. £	Plant £	Total £
T2538	NATURAL AND FAN ASSISTED CONVECTORS : remove convector access plate, remove convector final connections, reposition casing including redrilling and plugging wall as necessary, refix final connections, refix access plate					
156	500 - 1800 mm Long	nr		138.92		138.92
T2539	NATURAL AND FAN ASSISTED CONVECTORS : remove damaged convector front plate, repair as necessary and refix					
156	500 - 1800 mm Long	nr		34.73		34.73
T2540	NATURAL AND FAN ASSISTED CONVECTORS : remove damaged convector front plate, fix only new plate					
156	500 - 1800 mm Long	nr		13.31		13.31
T2541	NATURAL AND FAN ASSISTED CONVECTORS : remove damaged loose convector grille, fix only new grille					
156	500 - 1800 mm Long	nr		11.58		11.58
T2542	NATURAL AND FAN ASSISTED CONVECTORS : remove fan convector access plate, remove damaged air filter, fix only new air filter, refix casing					
156	500 - 1800 mm Long	nr		34.73		34.73
T2543	NATURAL AND FAN ASSISTED CONVECTORS : remove fan convector access plate, remove damaged or faulty on / off switch, speed controller, thermostat, audio thermostat					
160	Fix only replacement unit, refix casing	nr		17.37		17.37
162	Supply and fix new control unit to fan convector	nr	25.00	5.79		30.79
T2544	NATURAL AND FAN ASSISTED CONVECTORS : remove fan convector access plate, remove damaged heater battery, fix only new battery, refix casing					
156	500 - 1800 mm Long	nr		26.05		26.05
T2545	NATURAL AND FAN ASSISTED CONVECTORS : remove damaged / leaking natural convector, remove and set aside for re-use air cock / eliminator, plug and tails, final connections, fix only new convector, refix mountings and final connections, wall / floor type					
150	500 - 800 mm Long	nr		138.92		138.92
152	900 - 1200 mm Long	nr		162.07		162.07
154	1300 - 1800 mm Long	nr		188.12		188.12

© NSR 01 Aug 2008 - 31 Jul 2009   T10

T : MECHANICAL SERVICES - MECHANICAL HEATING/COOLING REFRIGERATION SYSTEMS : WORKS OF ALTERATION/SMALL WORKS/REPAIR			Mat. £	Lab. £	Plant £	Total £
T2546	NATURAL AND FAN ASSISTED CONVECTORS : remove damaged / leaking natural convector, remove and set aside for re-use air cock / eliminator, plug and tails, final connections, supply and fix new convector, refix mountings and final connections, wall / floor type					
150	500 - 800 mm Long	nr	270.00	138.92		408.92
152	900 - 1200 mm Long	nr	350.00	162.07		512.07
154	1300 - 1800 mm Long	nr	400.00	188.12		588.12
T2547	NATURAL AND FAN ASSISTED CONVECTORS : remove damaged / leaking natural convector, remove and set aside for re-use air cock / eliminator, plug and tails, final connections, fix only new convector, refix mountings and final connections, concealed type					
150	500 - 800 mm Long	nr		156.29		156.29
152	900 - 1200 mm Long	nr		179.44		179.44
154	1300 - 1800 mm Long	nr		205.49		205.49
T2548	NATURAL CONVECTORS : supply only natural convector, concealed type, up to 900 mm high					
150	500 - 800 mm Long	nr	305.00			305.00
152	900 - 1200 mm Long	nr	350.00			350.00
154	1300 - 1800 mm Long	nr	400.00			400.00
T2549	NATURAL AND FAN ASSISTED CONVECTORS : remove damaged / leaking natural convector, remove and set aside for re-use air cock / eliminator, plug and tails, final connections, supply and fix new convector, refix mountings and final connections, concealed type					
150	500 - 800 mm Long	nr	270.00	138.92		408.92
152	900 - 1200 mm Long	nr	350.00	138.92		488.92
154	1300 - 1800 mm Long	nr	400.00	138.92		538.92
T2550	NATURAL AND FAN ASSISTED CONVECTORS : remove damaged / leaking fan convector, remove and set aside for re-use air cock / eliminator, plug and tails, final connections, fix only new convector, refix mountings and final connections, wall floor type					
150	500 - 800 mm Long	nr		138.92		138.92
152	900 - 1200 mm Long	nr		162.07		162.07
154	1300 - 1800 mm Long	nr		188.12		188.12

© NSR 01 Aug 2008 - 31 Jul 2009

T11

T : MECHANICAL SERVICES - MECHANICAL HEATING/COOLING REFRIGERATION SYSTEMS : WORKS OF ALTERATION/SMALL WORKS/REPAIR			Mat. £	Lab. £	Plant £	Total £
T2551	NATURAL AND FAN ASSISTED CONVECTORS : remove damaged / leaking fan convector, remove and set aside for re-use air cock / eliminator, plug and tails, final connections, fix only convector, refix mountings and final connections, concealed type					
150	500 - 800 mm Long	nr		156.29		156.29
152	900 - 1200 mm Long	nr		179.44		179.44
154	1300 - 1800 mm Long	nr		205.49		205.49
T2552	NATURAL AND FAN ASSISTED CONVECTORS : supply only wall / floor mounted natural convector complete with brackets, air cock / eliminator and plug					
150	500 - 800 mm Long	nr	270.00			270.00
152	900 - 1200 mm Long	nr	350.00			350.00
154	1300 - 1800 mm Long	nr	400.00			400.00
T2553	NATURAL AND FAN ASSISTED CONVECTORS : remove damaged/leaking fan convector, remove and set aside for re-use air cock, plug and tails, final connections, supply and fix new convector, refix mountings and final connections, concealed type					
150	500 - 800 mm Long	nr	875.01	156.29		1031.30
152	900 - 1200 mm Long	nr	998.58	179.44		1178.02
154	1300 - 1800 mm long	nr	1237.63	205.49		1443.12
T2554	NATURAL CONVECTORS : fix only natural convector, wall/floor type, up to 900 mm high					
157	500 - 1800 mm Long	nr		82.77		82.77
T2555	NATURAL CONVECTORS : supply only natural convector, wall/floor type, up to 900 mm high					
150	500 - 800 mm Long	nr	270.00			270.00
152	900 - 1200 mm Long	nr	350.00			350.00
154	1300 - 1800 mm Long	nr	400.00			400.00
T2556	NATURAL CONVECTORS : supply and fix natural convector, wall/floor type, up to 900 mm high					
150	500 - 800 mm long	nr	270.00	82.77		352.77
152	900 - 1200 mm long	nr	350.00	88.56		438.56
154	1300 - 1800 mm long	nr	400.00	91.46		491.46

© NSR 01 Aug 2008 - 31 Jul 2009

T12

T : MECHANICAL SERVICES - MECHANICAL HEATING/COOLING REFRIGERATION SYSTEMS : WORKS OF ALTERATION/SMALL WORKS/REPAIR			Mat. £	Lab. £	Plant £	Total £
T2557	NATURAL CONVECTORS : fix only natural convector, concealed type, up to 900 mm high					
157	500 - 1800 mm Long	nr		98.40		98.40
T2558	NATURAL CONVECTORS : supply and fix natural convector, concealed type, up to 900 mm high					
150	500 - 800 mm long	nr	305.00	98.40		403.40
152	900 - 1200 mm long	nr	350.00	107.08		457.08
154	1300 - 1800 mm long	nr	400.00	112.87		512.87
T2559	FAN ASSISTED CONVECTORS : fix only fan convector, wall/floor type, up to 900 mm high					
176	900 - 1500 mm Long	nr		82.77		82.77
T2560	NATURAL AND FAN ASSISTED CONVECTORS : supply only concealed type natural convector complete with brackets, air-cock / eliminator and plug					
150	500 - 800 mm Long	nr	305.00			305.00
152	900 - 1200 mm Long	nr	350.00			350.00
154	1300 - 1800 mm Long	nr	400.00			400.00
T2561	NATURAL AND FAN ASSISTED CONVECTORS : supply only wall / floor mounted fan convector complete with brackets, air-cock / eliminator and plug					
150	500 - 800 mm Long	nr	875.01			875.01
152	900 - 1200 mm Long	nr	998.58			998.58
154	1300 - 1800 mm Long	nr	1237.63			1237.63
T2562	NATURAL AND FAN ASSISTED CONVECTORS : supply only concealed type fan convector complete with brackets air-cock / eliminator and plug					
150	500 - 800 mm Long	nr	838.42			838.42
152	900 - 1200 mm Long	nr	1024.06			1024.06
154	1300 - 1800 mm Long	nr	1263.58			1263.58
T2563	FAN ASSISTED CONVECTORS : supply and fix fan convector, ceiling type, up to 900 mm width					
170	900 mm long	nr	833.45	95.51		928.96
172	1200 mm long	nr	959.57	95.51		1055.08
174	1500 mm long	nr	1042.88	95.51		1138.39

© NSR 01 Aug 2008 - 31 Jul 2009

T13

T : MECHANICAL SERVICES - MECHANICAL HEATING/COOLING REFRIGERATION SYSTEMS : WORKS OF ALTERATION/SMALL WORKS/REPAIR				Mat. £	Lab. £	Plant £	Total £
T2564	FAN ASSISTED CONVECTORS : supply only fan convector, wall/floor type, up to 900 mm high						
	170	900 mm Long	nr	875.01			875.01
	172	1200 mm Long	nr	998.58			998.58
	174	1500 mm Long	nr	1237.63			1237.63
T2565	FAN ASSISTED CONVECTORS : supply and fix fan convector, wall/floor type, up to 900 mm high						
	170	900 mm long	nr	875.01	82.77		957.78
	172	1200 mm long	nr	998.58	82.77		1081.35
	174	1500 mm long	nr	1237.63	82.77		1320.40
T2566	FAN ASSISTED CONVECTORS : fix only fan convector, concealed type, up to 900 mm high						
	176	900 - 1500 mm Long	nr		86.83		86.83
T2567	FAN ASSISTED CONVECTORS : supply only fan convector, concealed type, up to 900 mm high						
	170	900 mm Long	nr	838.42			838.42
	172	1200 mm Long	nr	1024.06			1024.06
	174	1500 mm Long	nr	1263.58			1263.58
T2568	FAN ASSISTED CONVECTORS : supply and fix fan convector, concealed type, up to 900 mm high						
	170	900 mm long	nr	838.42	86.83		925.25
	172	1200 mm long	nr	1024.06	86.83		1110.89
	174	1500 mm long	nr	1263.58	86.83		1350.41
T2569	FAN ASSISTED CONVECTORS : fix only fan convector, concealed type, over 900 mm high						
	176	900 - 1500 mm Long	nr		115.77		115.77
T2570	FAN ASSISTED CONVECTORS : supply only fan convector, concealed type, over 900 mm high						
	170	900 mm Long	nr	807.25			807.25
	172	1200 mm Long	nr	1109.00			1109.00
	174	1500 mm Long	nr	1361.79			1361.79
T2571	FAN ASSISTED CONVECTORS : supply and fix fan convector, concealed type, over 900 mm high						
	170	900 mm long	nr	807.25	115.77		923.02

© NSR 01 Aug 2008 - 31 Jul 2009

T14

T : MECHANICAL SERVICES - MECHANICAL HEATING/COOLING REFRIGERATION SYSTEMS : WORKS OF ALTERATION/SMALL WORKS/REPAIR			Mat. £	Lab. £	Plant £	Total £
T2571						
172	1200 mm long	nr	1109.00	115.77		1224.77
174	1500 mm long	nr	1361.79	115.77		1477.56
T2572	FAN ASSISTED CONVECTORS : fix only fan convector, ceiling type, up to 900 mm width					
176	900 - 1500 mm Long	nr		95.51		95.51
T2573	FAN ASSISTED CONVECTORS : supply only fan convector, ceiling type, up to 900 mm width					
170	900 mm Long	nr	833.45			833.45
172	1200 mm Long	nr	959.57			959.57
174	1500 mm Long	nr	1042.88			1042.88
T2574	PERIMETER HEATING : supply only finned copper tube element section for perimeter heating					
230	500 mm Long, 15 mm diameter	nr	8.11			8.11
240	1080 mm long, 22 mm diameter	nr	55.65			55.65
246	2080 mm Long, 22 mm diameter, heavy duty	nr	154.35			154.35
256	1080 mm long, 28 mm diameter	nr	61.95			61.95
258	2080 mm Long, 28 mm diameter	nr	119.70			119.70
260	2080 mm Long, 28 mm diameter heavy duty	nr	158.55			158.55
T2575	PERIMETER HEATING : remove perimeter heating front plate, remove damaged / leaking finned copper tube element section, supply and fix new element including welding / jointing as necessary, refix front plate					
230	500 mm Long, 15 mm diameter	nr	8.11	20.26		28.37
240	1080 mm long, 22 mm diameter	nr	55.65	26.05		81.70
246	2080 mm long, 22 mm diameter, heavy duty	nr	154.35	37.62		191.97
250	2080 mm long, 28 mm diameter	nr	119.70	31.84		151.54
252	2080 mm long, 28 mm diameter heavy duty	nr	158.55	39.94		198.49
256	1080 mm long, 28 mm diameter	nr	61.95	28.94		90.89
260	1080 x 28 mm diameter, heavy duty	nr	158.55	31.84		190.39
T2576	PERIMETER HEATING : supply only finned tube mild steel element section for perimeter heating					
280	Up to 600 mm long, 25 mm diameter	nr	18.47			18.47
282	900 mm Long, 25 mm diameter	nr	27.72			27.72

© NSR 01 Aug 2008 - 31 Jul 2009   T15

T : MECHANICAL SERVICES - MECHANICAL HEATING/COOLING REFRIGERATION SYSTEMS : WORKS OF ALTERATION/SMALL WORKS/REPAIR			Mat. £	Lab. £	Plant £	Total £
T2576						
284	1200 mm Long, 25 mm diameter	nr	36.95			36.95
286	1800 mm Long, 25 mm diameter	nr	55.46			55.46
288	Up to 600 mm long, 32 mm diameter	nr	31.39			31.39
290	900 mm Long, 32 mm diameter	nr	46.98			46.98
292	1200 mm Long, 32 mm diameter	nr	62.75			62.75
294	1800 mm Long, 32 mm diameter	nr	94.12			94.12
T2577	PERIMETER HEATING : supply and fix finned tube mild steel element section for perimeter heating					
280	Up to 600 mm long, 25 mm diamete	nr	18.47	31.84		50.31
282	900 mm long, 25 mm diameter	nr	27.72	34.73		62.45
284	1200 mm long, 25 mm diameter	nr	36.95	37.62		74.57
286	1800 mm long, 25 mm diameter	nr	55.46	43.41		98.87
288	Up to 600 mm long, 32 mm diameter	nr	31.39	37.62		69.01
290	900 mm long, 32 mm diameter	nr	46.98	35.89		82.87
292	1200 mm long, 32 mm diameter	nr	62.75	42.25		105.00
294	1800 mm long, 32 mm diameter	nr	94.12	47.46		141.58
T2578	PERIMETER HEATING : fix only finned tube heating element, supports, including joints/couplers, single row					
200	22 mm Diameter, copper	m		24.31		24.31
202	28 mm Diameter, copper	m		31.84		31.84
204	35 mm Diameter, copper	m		37.05		37.05
206	25 mm Diameter, mild steel	m		41.68		41.68
208	32 mm Diameter, mild steel	m		48.62		48.62
T2579	PERIMETER HEATING : supply only finned tube heating element, supports, including joints/couplers, single row					
200	22 mm Diameter, copper	m	79.80			79.80
202	28 mm Diameter, copper	m	90.56			90.56
204	35 mm Diameter, copper	m	223.72			223.72
206	25 mm Diameter, mild steel	m	30.21			30.21
208	32 mm Diameter, mild steel	m	51.26			51.26

© NSR 01 Aug 2008 - 31 Jul 2009   T16

T : MECHANICAL SERVICES - MECHANICAL HEATING/COOLING REFRIGERATION SYSTEMS : WORKS OF ALTERATION/SMALL WORKS/REPAIR			Mat. £	Lab. £	Plant £	Total £
T2580	PERIMETER HEATING : supply and fix finned tube heating element, supports, including joints/couplers, single row					
200	22 mm diameter, copper	m	79.80	13.72		93.52
202	28 mm diameter, copper	m	90.56	18.41		108.97
204	35 mm diameter, copper	m	223.72	21.42		245.14
206	25 mm diameter, mild steel	m	30.21	24.10		54.31
208	32 mm diameter, mild steel	m	51.26	28.11		79.37
T2581	PERIMETER HEATING : fix only perimeter heating casing, including frontplate, grilles, enclosure pieces, dampers					
230	Any height/type casing	nr		52.10		52.10
232	Valve boxes	nr		13.39		13.39
234	Corner pieces	nr		20.08		20.08
236	End caps	nr		4.02		4.02
T2582	PERIMETER HEATING : supply only perimeter heating casing, including frontplate, grilles, enclosure pieces, dampers					
230	Any height/type casing	nr	207.94			207.94
232	Valve boxes	nr	67.31			67.31
234	Corner pieces	nr	51.10			51.10
236	End caps	nr	106.99			106.99
T2583	PERIMETER HEATING : supply and fix perimeter heating casing, including frontplate, grilles, enclosure pieces, dampers					
230	Any height/type casing	nr	207.94	52.10		260.04
232	Valve boxes	nr	67.31	13.39		80.70
234	Corner pieces	nr	51.10	20.08		71.18
236	End caps	nr	106.99	4.02		111.01
T2584	PERIMETER HEATING : remove damaged perimeter heating damper control mechanism and replace with new					
200	Supply and fix new mechanism	nr	21.02	12.16		33.18
T2585	PERIMETER HEATING : remove perimeter heating front plate, remove damaged casing joiner strip, fix only new joiner strip, refix front plate					
202	Supply and fix new joiner strip	nr	2.69	11.58		14.27

© NSR 01 Aug 2008 - 31 Jul 2009

T17

	T : MECHANICAL SERVICES - MECHANICAL HEATING/COOLING REFRIGERATION SYSTEMS : WORKS OF ALTERATION/SMALL WORKS/REPAIR		Mat. £	Lab. £	Plant £	Total £
T2586	PERIMETER HEATING : remove damaged perimeter heating front plate, repair as necessary and refix					
204	900 - 1800 mm Long	nr		34.73		34.73
T2587	PERIMETER HEATING : remove damaged perimeter heating front plate, fix only new front plate					
204	900 - 1800 mm Long	nr		13.31		13.31
T2588	PERIMETER HEATING : remove damaged perimeter heating top outlet grille, fix only new linear grille					
204	900 - 1800 mm Long	nr		11.58		11.58
T2589	PERIMETER HEATING : remove perimeter heating front plate, remove damaged / leaking finned copper tube element section, fix only new element including welding / jointing as necessary, refix front plate					
270	Up to 1800 mm long, 22 - 32 mm diameter	nr		26.05		26.05
T2590	PERIMETER HEATING : remove perimeter heating front plate, remove damaged / leaking finned tube mild steel element section, fix only new element including welding / jointing as necessary, refix front plate					
300	Up to 1800 mm long, 25 - 35 mm diameter	nr		31.84		31.84
T2591	PERIMETER HEATING : remove perimeter heating front plate, remove damaged / leaking expansion bellows, fix only new bellows, refix front plate					
310	22 mm Diameter	nr		23.15		23.15
312	28 mm Diameter	nr		28.94		28.94
314	35 mm Diameter	nr		34.73		34.73
T2592	PERIMETER HEATING : supply only expansion bellows for perimeter heating					
310	22 mm Diameter	nr	89.00			89.00
312	28 mm Diameter	nr	88.00			88.00
314	35 mm Diameter	nr	88.00			88.00
T2593	PERIMETER HEATING : remove perimeter heating front plate, remove damaged/leaking expansion bellows, supply and fix new, refix front plate					
310	22 mm Diameter	nr	89.00	23.15		112.15
312	28 mm Diameter	nr	88.00	28.94		116.94
314	35 mm Diameter	nr	88.00	34.73		122.73

© NSR 01 Aug 2008 - 31 Jul 2009 — T18

T : MECHANICAL SERVICES - MECHANICAL HEATING/COOLING REFRIGERATION SYSTEMS : WORKS OF ALTERATION/SMALL WORKS/REPAIR			Mat. £	Lab. £	Plant £	Total £
T2594	UNIT HEATERS : fix only unit heater, supports, horizontal, downward or angle discharge					
500	Wall mounted type	nr		173.65		173.65
502	Ceiling mounted type	nr		208.38		208.38
T2595	UNIT HEATERS : supply only unit heater, supports, horizontal, downward or angle discharge					
500	Wall mounted type	nr	807.48			807.48
502	Ceiling mounted type	nr	1419.46			1419.46
T2596	UNIT HEATERS : supply and fix unit heater, supports, horizontal, downward or angle discharge					
500	Wall mounted type	nr	807.48	173.65		981.13
502	Ceiling mounted type	nr	1419.46	208.38		1627.84
T2597	RADIANT PANELS : fix only radiant panel, concealed tubes, including joints/couplers, insulated, closure panels, supports.					
620	Single tube, average 2000 mm long, 160mm wide	nr		104.19		104.19
622	Single tube, average 4000 mm long, 160mm wide	nr		208.38		208.38
624	Single tube, average 6000 mm long, 160mm wide	nr		312.57		312.57
626	Double tube, average 2000 mm long, 320mm wide	nr		208.38		208.38
628	Double tube, average 4000 mm long, 320mm wide	nr		416.76		416.76
630	Double tube, average 6000 mm long, 320mm wide	nr		625.14		625.14
632	Triple tube, average 2000 mm long, 480mm wide	nr		312.57		312.57
634	Triple tube, average 4000 mm long, 480mm wide	nr		625.14		625.14
636	Triple tube, average 6000 mm long, 480mm wide	nr		937.71		937.71
T2598	RADIANT PANELS : supply only radiant panel, concealed tubes, including joints/couplers, insulated, closure panels, supports					
620	Single tube, average 2000 mm long, 160mm wide	nr	122.00			122.00
622	Single tube, average 4000 mm long, 160mm wide	nr	144.00			144.00
624	Single tube, average 6000 mm long, 160mm wide	nr	201.00			201.00
626	Double tube, average 2000 mm long, 320mm wide	nr	132.00			132.00
628	Double tube, average 4000 mm long, 320mm wide	nr	267.00			267.00
630	Double tube, average 6000 mm long, 320mm wide	nr	382.00			382.00
632	Triple tube, average 2000 mm long, 480mm wide	nr	193.00			193.00
634	Triple tube, average 4000 mm long, 480mm wide	nr	391.00			391.00
636	Triple tube, average 6000 mm long, 480mm wide	nr	562.00			562.00

© NSR 01 Aug 2008 - 31 Jul 2009   T19

T : MECHANICAL SERVICES - MECHANICAL HEATING/COOLING REFRIGERATION SYSTEMS : WORKS OF ALTERATION/SMALL WORKS/REPAIR			Mat. £	Lab. £	Plant £	Total £
T2599	RADIANT PANELS : supply and fix radiant panel, concealed tubes, including joints/couplers, insulated, closure panels, supports.					
620	Single tube, average 2000 mm long, 160 mm wide	nr	122.00	104.19		226.19
622	Single tube, average 4000 mm long, 160 mm wide	nr	144.00	208.38		352.38
624	Single tube, average 6000 mm long, 160 mm wide	nr	201.00	312.57		513.57
626	Double tube, average 2000 mm long, 320 mm wide	nr	132.00	208.38		340.38
628	Double tube, average 4000 mm long, 320 mm wide	nr	267.00	416.76		683.76
630	Double tube, average 6000 mm long, 320 mm wide	nr	382.00	625.14		1007.14
632	Triple tube, average 2000 mm long, 480 mm wide	nr	193.00	312.57		505.57
634	Triple tube, average 4000 mm long, 480 mm wide	nr	391.00	625.14		1016.14
636	Triple tube, average 6000 mm long, 480 mm wide	nr	562.00	937.71		1499.71
	T27 : SPACE HEATING SERVICES : PIPEWORK AND ANCILLARIES					
T2700	GENERAL : supply and fix braided rubber hose assembly, with threaded end fittings					
044	15 mm Diameter, 300 mm long	nr	6.13	10.04		16.17
046	20 mm Diameter, 300 mm long	nr	8.18	11.71		19.89
048	15 mm Diameter, 600 mm long	nr	7.91	13.05		20.96
050	20 mm Diameter, 600 mm long	nr	11.03	15.06		26.09
T2704	GENERAL : fix only rubber compensator					
031	15 - 20 mm Diameter	nr		25.10		25.10
033	25 - 32 mm Diameter	nr		37.48		37.48
035	40 - 50 mm Diameter	nr		50.20		50.20
037	65 - 80 mm Diameter	nr		75.30		75.30
039	100 mm Diameter	nr		100.40		100.40
T2708	GENERAL : remove damaged rubber compensator, fix only new compensator					
030	15 - 20 mm Diameter	nr		28.94		28.94
032	25 - 32 mm Diameter	nr		43.41		43.41
034	40 - 50 mm Diameter	nr		52.10		52.10
036	65 - 80 mm Diameter	nr		69.46		69.46
038	100 mm Diameter	nr		86.83		86.83

© NSR 01 Aug 2008 - 31 Jul 2009

T20

T : MECHANICAL SERVICES - MECHANICAL HEATING/COOLING REFRIGERATION SYSTEMS : WORKS OF ALTERATION/SMALL WORKS/REPAIR			Mat. £	Lab. £	Plant £	Total £
T2712	GENERAL : remove damaged rubber compensator, supply and fix new compensator					
030	15 - 20 mm Diameter	nr	21.00	57.88		78.88
032	25 - 32 mm Diameter	nr	30.00	43.41		73.41
034	40 - 50 mm Diameter	nr	48.00	52.10		100.10
036	65 - 80 mm Diameter	nr	93.00	69.46		162.46
038	100 mm Diameter	nr	160.00	86.83		246.83
T2716	GENERAL : supply only rubber compensator					
030	15 - 20 mm Diameter	nr	21.00			21.00
032	25 - 32 mm Diameter	nr	26.00			26.00
034	40 - 50 mm Diameter	nr	48.00			48.00
036	65 - 80 mm Diameter	nr	93.00			93.00
038	100 mm Diameter	nr	167.00			167.00
T2720	GENERAL : supply and fix rubber compensator					
030	15 - 20 mm diameter	nr	21.00	25.10		46.10
032	25 - 32 mm diameter	nr	30.00	37.48		67.48
034	40 - 50 mm diameter	nr	48.00	50.20		98.20
036	65 - 80 mm diameter	nr	93.00	75.30		168.30
038	100 mm diameter	nr	167.00	100.40		267.40
T2724	GENERAL : remove damaged expansion bellows, fix only new bellows					
050	15 - 20 mm Diameter	nr		12.21		12.21
052	25 - 32 mm Diameter	nr		25.10		25.10
054	40 - 50 mm Diameter	nr		33.47		33.47
056	65 - 80 mm Diameter	nr		50.20		50.20
058	100 mm Diameter	nr		66.93		66.93
T2728	GENERAL : remove damaged expansion bellows, supply and fix new bellows					
050	15 - 20 mm Diameter	nr	147.00	16.73		163.73
052	25 - 32 mm Diameter	nr	164.00	25.10		189.10
054	40 - 50 mm Diameter	nr	169.00	33.47		202.47
056	65 - 80 mm Diameter	nr	192.00	50.20		242.20
058	100 mm Diameter	nr	238.00	66.93		304.93

© NSR 01 Aug 2008 - 31 Jul 2009

T21

T : MECHANICAL SERVICES - MECHANICAL HEATING/COOLING REFRIGERATION SYSTEMS : WORKS OF ALTERATION/SMALL WORKS/REPAIR			Mat. £	Lab. £	Plant £	Total £
T2736	GENERAL : fix only expansion bellows					
044	15 mm Diameter	nr		23.43		23.43
046	20 mm Diameter	nr		26.77		26.77
048	25 mm Diameter	nr		33.47		33.47
051	32 mm Diameter	nr		41.83		41.83
053	40 mm Diameter	nr		46.85		46.85
055	50 mm Diameter	nr		53.55		53.55
057	65 mm Diameter	nr		66.93		66.93
059	80 mm Diameter	nr		83.67		83.67
060	100 mm Diameter	nr		100.40		100.40
T2740	GENERAL : supply only expansion bellows					
044	15 mm Diameter	nr	66.00			66.00
046	20 mm Diameter	nr	139.00			139.00
048	25 mm Diameter	nr	140.00			140.00
051	32 mm Diameter	nr	164.00			164.00
053	40 mm Diameter	nr	169.00			169.00
055	50 mm Diameter	nr	142.00			142.00
057	65 mm Diameter	nr	164.00			164.00
059	80 mm Diameter	nr	192.00			192.00
061	100 mm Diameter	nr	238.00			238.00
T2744	GENERAL : supply and fix expansion bellows					
044	15 mm diameter	nr	66.00	23.43		89.43
046	20 mm diameter	nr	139.00	26.77		165.77
048	25 mm diameter	nr	140.00	33.47		173.47
050	32 mm diameter	nr	164.00	41.83		205.83
052	40 mm diameter	nr	169.00	46.85		215.85
054	50 mm diameter	nr	142.00	53.55		195.55
056	65 mm diameter	nr	164.00	66.93		230.93
058	80 mm diameter	nr	192.00	83.67		275.67
060	100 mm diameter	nr	238.00	100.40		338.40
T2748	RADIATOR ANCILLARIES					
436	Remove damaged radiator air cock, supply and fix new 12 mm diameter	nr	1.68	6.69		8.37

© NSR 01 Aug 2008 - 31 Jul 2009   T22

T : MECHANICAL SERVICES - MECHANICAL HEATING/COOLING REFRIGERATION SYSTEMS : WORKS OF ALTERATION/SMALL WORKS/REPAIR			Mat. £	Lab. £	Plant £	Total £
T2748						
438	Remove damaged radiator valve, hand wheel or lockshield, supply and fix new 15 - 25 mm diameter	nr	13.50	11.38		24.88
T2752	**RADIATOR ANCILLARIES : supply and fit valve and components**					
500	Thermostatic radiator valve, 15 mm	nr	13.50	7.36		20.86
501	Thermostatic radiator valve, 20 mm	nr	54.16	8.70		62.86
502	Thermostatic radiator valve, 25 mm	nr	23.06	10.71		33.77
T2756	**RADIATOR ANCILLARIES : valve and components**					
440	Remove damaged radiator wheel or lockshield valve, supply and fix new, reset valve position 15 mm diameter	nr	31.64	7.70		39.34
442	Remove damaged radiator wheel or lockshield valve, supply and fix new, reset valve position 20 mm diameter	nr	24.50	8.70		33.20
444	Remove damaged radiator wheel or lockshield valve, supply and fix new, reset valve position 25 mm diameter	nr	39.27	10.37		49.64
446	Remove thermostatic radiator valve gland seal, supply and fix new, reset valve position 15 - 25 mm diameter	nr	6.86	6.69		13.55
448	Remove damaged thermostatic radiator valve in-built sensor head, supply and fix new, reset 15 - 25 mm diameter	nr	26.19	8.03		34.22
450	Remove damaged thermostatic radiator valve remote sensor and capillary, supply and fix new, reset 15 - 25 mm diameter	nr	30.22	11.71		41.93
T2760	**RADIATOR ANCILLARIES : remove damaged thermostatic valve body, supply and fix new**					
460	15 mm Diameter	nr	33.94	7.03		40.97
462	20 mm Diameter	nr	54.65	8.03		62.68
464	25 mm Diameter	nr	49.43	9.37		58.80
T2764	**RADIATOR ANCILLARIES : remove damaged thermostatic radiator valve and in-built sensor, supply and fix new, reset valve position**					
460	15 mm Diameter	nr	23.50	10.04		33.54
462	20 mm Diameter	nr	37.27	11.71		48.98
464	25 mm Diameter	nr	69.68	13.05		82.73
T2768	**RADIATOR ANCILLARIES : remove damaged thermostatic radiator valve, remote sensor capillary, supply and fix new, reset valve position**					
460	15 mm Diameter	nr	28.75	14.06		42.81
462	20 mm Diameter	nr	33.87	15.73		49.60

© NSR 01 Aug 2008 - 31 Jul 2009   T23

T : MECHANICAL SERVICES - MECHANICAL HEATING/COOLING REFRIGERATION SYSTEMS : WORKS OF ALTERATION/SMALL WORKS/REPAIR			Mat. £	Lab. £	Plant £	Total £
T2768						
464	25 mm Diameter	nr	33.87	17.07		50.94
T2772	**RADIATOR ANCILLARIES : remove damaged manual radiator valve, supply and fix new thermostatic radiator valve and in-built sensor, set valve position**					
460	15 mm Diameter	nr	23.50	10.04		33.54
462	20 mm Diameter	nr	37.27	11.71		48.98
464	25 mm Diameter	nr	69.68	13.05		82.73
T2776	**RADIATOR ANCILLARIES : remove damaged manual radiator valve, supply and fix new thermostatic radiator valve, remote sensor and capillary, set valve position**					
460	15 mm Diameter	nr	93.55	14.06		107.61
462	20 mm Diameter	nr	96.04	15.73		111.77
464	25 mm Diameter	nr	107.16	17.07		124.23
T2780	**NATURAL AND FAN ASSISTED CONVECTOR ANCILLARIES**					
480	Remove convector casing as necessary, remove damaged manual air cock, supply and fix new, refix casing 6 mm diameter	nr	1.68	53.25		54.93
484	Remove convector casing as necessary, remove damaged automatic air vent, supply and fix new, refix casing 6 mm diameter	nr	5.49	55.57		61.06
486	Remove convector access plate, tighten or repack convector valve gland, grease spindle, refix casing 15 - 25 mm diameter	nr		66.93		66.93
490	Remove convector access plate, remove damaged wheel or lockshield valve, supply and fix new, reset valve position, 15 - 25 mm diameter	nr	4.81	64.26		69.07
T2784	**NATURAL AND FAN ASSISTED CONVECTOR ANCILLARIES : remove damaged convector wheel or lockshield valve, supply and fix new, reset valve position**					
460	15 mm Diameter	nr	17.13	70.28		87.41
462	20 mm Diameter	nr	24.21	71.95		96.16
464	25 mm Diameter	nr	27.64	73.63		101.27
T2788	**GENERALLY**					
950	Draining down domestic heating system, including F&E tank, and leaving in a safe condition	nr		80.32		80.32
952	Draining down domestic heating system, including F&E tank, in order to carry out alterations/repairs	nr		80.32		80.32
954	Filling up domestic heating system, including venting & the like	nr		80.32		80.32

© NSR 01 Aug 2008 - 31 Jul 2009   T24

T : MECHANICAL SERVICES - MECHANICAL HEATING/COOLING REFRIGERATION SYSTEMS : WORKS OF ALTERATION/SMALL WORKS/REPAIR			Mat. £	Lab. £	Plant £	Total £
T2789	DRAINING DOWN OR REFILLING OF NON-DOMESTIC HEATING SYSTEM - zoned or small building (area based on zone)					
950	Up to 200 m2	m2		0.37		0.37
952	From 200 - 1500 m2	m2		0.18		0.18
954	1500 - 4000 m2	m2		0.17		0.17
956	4000 and upwards	m2		0.15		0.15
	T28 : SPACE HEATING SERVICES: PROGRAMMERS AND TIMERS					
T2810	PROGRAMMERS AND TIMERS (SANGAMO): Supply only programmer/timer					
050	Compact quartz time switch single channel 24hr	nr	30.57			30.57
060	Compact quartz time switch single channel 7 day	nr	38.16			38.16
100	Electronic timer 3pin 7 day	nr	171.57			171.57
110	Electronic timer 4pin 7 day	nr	174.45			174.45
120	Q554 Form 2 Time Switch 3 pin 24hr	nr	125.38			125.38
130	Q555 Form 2 Time Switch 4 pin 24hr	nr	130.29			130.29
T2815	PROGRAMMERS AND TIMERS (SANGAMO): Supply and fix wall mounted, including fixings, connections					
050	Compact quartz time switch single channel 24hr	nr	30.57	24.53		55.10
060	Compact quartz time switch single channel 7 day	nr	38.16	24.53		62.69
100	Electronic timer 3pin 7 day	nr	171.57	28.34		199.91
110	Electronic timer 4pin 7 day	nr	174.45	28.34		202.79
120	Q554 Form 2 Time Switch 3 pin 24hr	nr	125.38	28.34		153.72
130	Q555 Form 2 Time Switch 4 pin 24hr	nr	130.29	28.34		158.63
T2820	PROGRAMMERS AND TIMERS (SANGAMO): Fix only wall mounted, including fixings, connections					
050	Compact quartz time switch single channel 24hr	nr		24.53		24.53
060	Compact quartz time switch single channel 7 day	nr		24.53		24.53
100	Electronic timer 3pin 7 day	nr		28.34		28.34
110	Electronic timer 4pin 7 day	nr		28.34		28.34
120	Q554 Form 2 Time Switch 3 pin 24hr	nr		28.34		28.34
130	Q555 Form 2 Time Switch 4 pin 24hr	nr	130.29	28.34		158.63

© NSR 01 Aug 2008 - 31 Jul 2009   T25

T : MECHANICAL SERVICES - MECHANICAL HEATING/COOLING REFRIGERATION SYSTEMS : WORKS OF ALTERATION/SMALL WORKS/REPAIR			Mat. £	Lab. £	Plant £	Total £
T2825	PROGRAMMERS AND TIMERS (SANGAMO): Remove damaged equipment, Supply and fix wall mounted, including fixings, connections					
050	Compact quartz time switch single channel 24hr	nr	30.57	28.76		59.33
060	Compact quartz time switch single channel 7 day	nr	38.16	28.76		66.92
100	Electronic timer 3pin 7 day	nr	171.57	32.57		204.14
110	Electronic timer 4pin 7 day	nr	174.45	32.57		207.02
120	Q554 Form 2 Time Switch 3 pin 24hr	nr	125.38	32.57		157.95
130	Q555 Form 2 Time Switch 4 pin 24hr	nr	130.29	32.57		162.86
	T55 : REFRIGERATION					
T5500	PIPEWORK : remove damaged copper pipework, BS2871, fittings, armaflex insulation, brackets, supply and install new					
010	1/4" (6 mm) diameter	m	11.92	12.10		24.02
011	3/8" (9 mm) diameter	m	12.91	12.96		25.87
012	1/2" (12.5 mm) diameter	m	25.71	14.40		40.11
013	5/8" (16 mm) diameter	m	32.58	16.13		48.71
014	3/4" (20 mm) diameter	m	32.58	17.28		49.86
017	1.1/8" (28 mm) diameter	m	38.30	20.74		59.04
T5506	PIPEWORK : remove damaged condensate pipework, PVC, vinyl braided type, brackets and fittings, supply and install new					
020	15 mm diameter	m	3.67	6.62		10.29
T5512	PIPEWORK : condensate, PVC, braided type, brackets and fittings					
021	15 mm diameter	m	7.01	4.32		11.33
T5518	PIPEWORK : supply and fit copper pipework (TX), EN1057 R250, capillary fittings, armaflex insulation, brackets					
020	1/4" diameter	m	8.69	8.06		16.75
021	3/8" diameter	m	10.55	8.64		19.19
022	1/2" diameter	m	10.40	9.50		19.90
023	5/8" diameter	m	10.44	10.66		21.10
024	3/4" diameter	m	17.68	11.52		29.20
027	1.1/8" diameter	m	28.49	13.54		42.03

T : MECHANICAL SERVICES - MECHANICAL HEATING/COOLING REFRIGERATION SYSTEMS : WORKS OF ALTERATION/SMALL WORKS/REPAIR			Mat. £	Lab. £	Plant £	Total £
T5524	REFRIGERANT GAS : remove refrigerant from circuit, re-charge re-using refrigerant, up to 20 metres between A/C unit and condensers					
105	R22	nr		207.36		207.36
T5530	REFRIGERANT GAS : supply and install additional refrigerant gas to air conditioning unit					
150	Type R22	kg	13.70	28.80		42.50
T5536	UNIT PARTS : supply only item of equipment					
150	Compressor, 5-6 kW	nr	410.00			410.00
250	Fan motor, 1 kw	nr	96.64			96.64
270	Impeller	nr	18.36			18.36
290	Fan Blade	nr	18.36			18.36
T5542	UNIT PARTS : supply and fix item of equipment, cassette or high wall type A/C unit.					
150	Compressor, 5-6 kW	nr	410.00	144.00		554.00
250	Fan motor, 1 kwt.	nr	96.64	69.12		165.76
270	Impeller	nr	18.36	51.84		70.20
290	Fan Blade	nr	18.36	17.28		35.64
T5548	UNIT PARTS : remove faulty or damaged item of equipment, install only new item of equipment, cassette or high wall type A/C unit					
150	Compressor, 5-6 kW	nr		144.00		144.00
190	PCB	nr		34.56		34.56
250	Fan motor, 1 kw	nr		69.12		69.12
270	Impeller	nr		51.84		51.84
290	Fan blade	nr		17.28		17.28
300	Compressor, evaporator, 4 way valve and TX valve assembly	nr		103.68		103.68
460	Expansion valve	nr		120.96		120.96
T5554	COMPONENT PARTS : remove faulty or damaged item of equipment, supply and install new, cassette or high wall type A/C unit					
500	Filter	nr	22.11	1.44		23.55
550	Capacitor, 50-60 mf	nr	11.84	17.28		29.12
600	Contactor, 4 kW	nr	14.74	17.28		32.02

© NSR 01 Aug 2008 - 31 Jul 2009

T : MECHANICAL SERVICES - MECHANICAL HEATING/COOLING REFRIGERATION SYSTEMS : WORKS OF ALTERATION/SMALL WORKS/REPAIR			Mat. £	Lab. £	Plant £	Total £
T5560	MAINTENANCE : carry out maintenance activities as directed					
700	Clean filters	nr		7.20		7.20
	T76 : GAS SERVICES					
T7600	GAS HEATERS					
010	Remove damaged valve assembly and regulator, fix only new assembly	nr		40.16		40.16
012	Supply and fix new gas heater valve assembly and regulator	nr	302.93	20.08		323.01
015	Remove damaged gas water heater swivel spout, fix only new spout	nr		14.06		14.06
020	Supply and fix new swivel spout	nr	12.21	7.03		19.24
025	Refix gas water heater including redrilling and plugging wall as necessary	nr		60.24		60.24
030	Remove damaged gas water heater, fix only new heater 7 - 15 litre capacity	nr		60.24		60.24
035	Supply only gas water heater 7 - 15 litre capacity	nr	178.75			178.75
S35	Remove damaged 7 - 15 litre capacity gas water heater, supply and fix gas water heater	nr	178.75	60.24		238.99
T7625	GAS HEATERS : fix only gas water heater					
011	7 - 15 Litre capacity	nr		15.59		15.59
T7650	GAS HEATERS : supply only gas water heater					
010	10 Litre capacity	nr	178.75			178.75
T7675	GAS HEATERS : supply and fix gas water heater					
010	10 Litre capacity	nr	178.75	15.59		194.34
	T80 : PLANT ROOM SERVICES					
T8000	CIRCULATING PUMPS					
010	Remove damaged vee belt, fix only new	nr		20.08		20.08
015	Supply and fix new vee belt	nr	30.96	10.04		41.00
T8002	CIRCULATING PUMPS : remove existing inline pipeline mounted circulating pump, fix only new					
020	20 - 32 mm Diameter line size	nr		69.46		69.46
022	40 - 50 mm Diameter line size	nr		86.83		86.83
024	65 mm Diameter line size	nr		104.19		104.19

© NSR 01 Aug 2008 - 31 Jul 2009   T28

T : MECHANICAL SERVICES - MECHANICAL HEATING/COOLING REFRIGERATION SYSTEMS : WORKS OF ALTERATION/SMALL WORKS/REPAIR			Mat. £	Lab. £	Plant £	Total £
T8004	CIRCULATING PUMPS : supply only inline pipeline mounted circulating pump					
020	20 - 32 mm Diameter line size	nr	169.48			169.48
022	40 - 50 mm Diameter line size	nr	409.00			409.00
024	65 mm Diameter line size	nr	727.00			727.00
T8006	CIRCULATING PUMPS : remove existing inline pipeline mounted circulating pump, supply and fix new					
020	20 - 32 mm diameter line size,	nr	169.48	69.46		238.94
022	40 - 50 mm diameter line size,	nr	409.00	86.83		495.83
024	65 mm diameter line size	nr	727.00	104.19		831.19
T8008	CIRCULATING PUMPS : remove existing inline pipeline mounted twin circulating pumps, fix only new					
018	32 - 50 mm Diameter line size	nr		92.61		92.61
024	65 mm Diameter line size	nr		109.98		109.98
026	80 mm Diameter line size	nr		127.34		127.34
T8010	CIRCULATING PUMPS : supply only inline pipeline mounted twin circulating pumps					
018	32 - 50 mm Diameter line size	nr	1519.00			1519.00
024	65 mm Diameter line size	nr	2234.00			2234.00
026	80 mm Diameter line size	nr	3795.00			3795.00
T8012	CIRCULATING PUMPS : remove existing inline pipeline mounted twin circulating pumps, supply and fix new					
018	32 - 50 mm diameter line size,	nr	1519.00	92.61		1611.61
024	65 mm diameter line size,	nr	2234.00	109.98		2343.98
026	80 mm diameter line size,	nr	3795.00	127.34		3922.34
T8014	CIRCULATING PUMPS : remove existing floor mounted circulating pump, fix only new					
020	20 - 32 mm Diameter line size	nr		75.25		75.25
022	40 - 50 mm Diameter line size	nr		92.61		92.61
024	65 mm Diameter line size	nr		109.98		109.98
026	80 mm Diameter line size	nr		121.56		121.56
028	100 mm Diameter line size	nr		138.92		138.92

© NSR 01 Aug 2008 - 31 Jul 2009   T29

T : MECHANICAL SERVICES - MECHANICAL HEATING/COOLING REFRIGERATION SYSTEMS : WORKS OF ALTERATION/SMALL WORKS/REPAIR				Mat. £	Lab. £	Plant £	Total £
T8016	CIRCULATING PUMPS : supply only floor mounted circulating pump						
020	20 - 32 mm Diameter line size		nr	548.95			548.95
022	40 - 50 mm Diameter line size		nr	1001.34			1001.34
024	65 mm Diameter line size		nr	1105.84			1105.84
026	80 mm Diameter line size		nr	1142.87			1142.87
028	100 mm Diameter line size		nr	1298.96			1298.96
T8018	CIRCULATING PUMPS : remove existing floor mounted circulating pump, supply and fix new						
020	20 - 32 diameter line size,		nr	548.95	75.25		624.20
022	40 - 50 diameter line size,		nr	1001.34	92.61		1093.95
024	65 diameter line size,		nr	1105.84	109.98		1215.82
026	80 diameter line size,		nr	1142.87	121.56		1264.43
028	100 diameter line size,		nr	1298.96	138.92		1437.88
T8020	CONTROL VALVES : remove existing motor assembly, fix only new						
110	15 - 50 mm Diameter line size		nr		18.07		18.07
112	65 - 100 mm Diameter line size		nr		20.08		20.08
T8022	CONTROL VALVES : supply only valve motor assembly						
110	15 - 50 mm Diameter line size (PC £200)		nr	200.00			200.00
112	65 - 100 mm Diameter line size (PC £300)		nr	300.00			300.00
T8024	CONTROL VALVES : remove existing valve motor assembly, supply and fix new						
110	15 - 50 mm diameter line size (PC £200)		nr	200.00	18.07		218.07
112	65 - 100 mm diameter line size (PC £300)		nr	300.00	20.08		320.08
T8026	CONTROL VALVES : remove existing two port control valve and motor, fix only new						
016	15 - 25 mm Diameter line size		nr		15.63		15.63
018	32 - 50 mm Diameter line size		nr		22.00		22.00
024	65 mm Diameter line size		nr		27.78		27.78
026	80 mm Diameter line size		nr		32.99		32.99
028	100 mm Diameter line size		nr		36.47		36.47

© NSR 01 Aug 2008 - 31 Jul 2009   T30

T : MECHANICAL SERVICES - MECHANICAL HEATING/COOLING REFRIGERATION SYSTEMS : WORKS OF ALTERATION/SMALL WORKS/REPAIR			Mat. £	Lab. £	Plant £	Total £
T8028	CONTROL VALVES : supply only two port control valve and motor					
016	15 - 25 mm Diameter line size	nr	275.00			275.00
018	32 - 50 mm Diameter line size	nr	339.00			339.00
024	65 mm Diameter line size	nr	559.00			559.00
026	80 mm Diameter line size	nr	747.00			747.00
028	100 mm Diameter line size	nr	1057.00			1057.00
T8030	CONTROL VALVES : remove existing two port control valve and motor, supply and fix new					
016	15 - 25 mm diameter line size,	nr	275.00	15.63		290.63
018	32 - 50 mm diameter line size,	nr	339.00	22.00		361.00
024	65 mm diameter line size,	nr	559.00	27.78		586.78
026	80 mm diameter line size	nr	747.00	32.99		779.99
028	100 mm diameter line size,	nr	1057.00	36.47		1093.47
T8032	CONTROL VALVES : remove existing three port control valve and motor, fix only new					
016	15 - 25 mm Diameter line size	nr		22.57		22.57
018	32 - 50 mm Diameter line size	nr		31.26		31.26
024	65 mm Diameter line size	nr		39.94		39.94
026	80 mm Diameter line size	nr		46.89		46.89
028	100 mm Diameter line size	nr		52.10		52.10
T8034	CONTROL VALVES : supply only three port control valve and motor					
016	15 - 25 mm Diameter line size	nr	318.00			318.00
018	32 - 50 mm Diameter line size	nr	360.00			360.00
024	65 mm Diameter line size	nr	556.00			556.00
026	80 mm Diameter line size	nr	648.00			648.00
028	100 mm Diameter line size	nr	900.00			900.00
T8036	CONTROL VALVES : remove existing three port control valve and motor, supply and fix new					
016	15 - 25 mm diameter line size,	nr	318.00	22.57		340.57
018	32 - 50 mm diameter line size,	nr	360.00	31.26		391.26
024	65 mm diameter line size,	nr	556.00	39.94		595.94
026	80 mm diameter line size	nr	648.00	46.89		694.89

© NSR 01 Aug 2008 - 31 Jul 2009

T31

T : MECHANICAL SERVICES - MECHANICAL HEATING/COOLING REFRIGERATION SYSTEMS : WORKS OF ALTERATION/SMALL WORKS/REPAIR			Mat. £	Lab. £	Plant £	Total £
T8036						
028	100 mm diameter line size,	nr	900.00	52.10		952.10
T8038	**GAS BURNERS AND GAS TRAINS : remove existing gas burner accessory, fix only new**					
600	Solenoid valve (up to 30 kW)	nr		12.16		12.16
640	Ignition probe (up to 30 kW)	nr		5.02		5.02
641	Ignition probe (up to 300 kW)	nr		6.69		6.69
660	Ignition sensor (up to 30 kW)	nr		5.02		5.02
661	Ignition sensor (up to 300 kW)	nr		6.69		6.69
675	Sensing electrode (up to 100 kW)	nr		15.06		15.06
720	Control box (up to 30 kW)	nr		6.69		6.69
722	Control box (over 30 kW)	nr		15.06		15.06
750	Burner motor (up to 20 kW)	nr		33.47		33.47
752	Burner motor (20 kW to 30 kW)	nr		40.16		40.16
754	Burner motor (30 kW to 300 kW)	nr		60.24		60.24
760	Burner motor (over 300 kW)	nr		80.32		80.32
800	Thermocouple	nr		5.02		5.02
820	Nozzle	nr		4.02		4.02
T8040	**GAS BURNERS AND GAS TRAINS : supply only gas burner accessories**					
600	Solenoid valve (up to 30 kW)	nr	64.83			64.83
640	Ignition probe (up to 30 kW)	nr	5.12			5.12
641	Ignition probe (up to 300 kW)	nr	8.76			8.76
660	Ignition sensor (up to 30 kW)	nr	5.12			5.12
661	Ignition sensor (up to 300 kW)	nr	8.76			8.76
675	Sensing electrode (up to 100 kW)	nr	27.90			27.90
720	Control box (up to 30 kW)	nr	52.58			52.58
722	Control box (over 30 kW)	nr	175.30			175.30
750	Burner motor (up to 20 kW)	nr	40.18			40.18
752	Burner motor (20 kW to 30 kW)	nr	43.83			43.83
754	Burner motor (30 kW to 300 kW)	nr	87.66			87.66
760	Burner motor (over 300 kW)	nr	233.74			233.74
800	Thermocouple	nr	5.20			5.20

© NSR 01 Aug 2008 - 31 Jul 2009

T32

T : MECHANICAL SERVICES - MECHANICAL HEATING/COOLING REFRIGERATION SYSTEMS : WORKS OF ALTERATION/SMALL WORKS/REPAIR			Mat. £	Lab. £	Plant £	Total £
T8040						
820	Nozzle	nr	6.26			6.26
T8042	**GAS BURNERS AND GAS TRAINS : remove existing gas burner accessory, supply and fix new**					
600	Solenoid valve (up to 30 kW)	nr	64.83	12.16		76.99
640	Ignition probe (up to 30 kW),	nr	5.12	5.02		10.14
641	Ignition probe (up to 300 kW),	nr	8.76	6.69		15.45
660	Ignition sensor (up to 30 kW),	nr	5.12	5.02		10.14
661	Ignition sensor (up to 300 kW),	nr	8.76	6.69		15.45
675	Sensing electrode (up to 100 kW),	nr	27.90	15.06		42.96
720	Control box (up to 30 kW)	nr	52.58	6.69		59.27
722	control box (over 30 kW),	nr	175.30	15.06		190.36
750	Burner motor (up to 20 kW)	nr	40.18	33.47		73.65
752	Burner motor (20 kW to 30 kW),	nr	43.83	40.16		83.99
754	Burner motor (30 kW to 300 kW),	nr	87.66	60.24		147.90
760	Burner motor (over 300 kW)	nr	233.74	80.32		314.06
800	Thermocouple	nr	5.20	5.02		10.22
820	Nozzle,	nr	6.26	4.02		10.28
T8044	**GAS BURNERS AND GAS TRAINS : remove existing gas burner and train, fix only new**					
232	Single stage burner, 25 - 60 kw	nr		86.83		86.83
234	Single stage burner, 60 - 250 kw	nr		104.19		104.19
236	Single stage burner, 250 - 380 kw	nr		138.92		138.92
238	Single stage burner, 380 - 600 kw	nr		173.65		173.65
T8046	**GAS BURNERS AND GAS TRAINS : supply only gas burner and train**					
232	Single stage burner, 25 - 60 kw	nr	468.96			468.96
234	Single stage burner, 60 - 250 kw	nr	794.72			794.72
236	Single stage burner, 250 - 380 kw	nr	1306.02			1306.02
238	Single stage burner, 380 - 600 kw	nr	4497.46			4497.46
T8048	**GAS BURNERS AND GAS TRAINS : remove existing gas burner and train, supply and fix new**					
232	Single stage burner, 25 - 60 kW,	nr	468.96	86.83		555.79

© NSR 01 Aug 2008 - 31 Jul 2009   T33

T : MECHANICAL SERVICES - MECHANICAL HEATING/COOLING REFRIGERATION SYSTEMS : WORKS OF ALTERATION/SMALL WORKS/REPAIR			Mat. £	Lab. £	Plant £	Total £
T8048						
234	Single stage burner, 60 - 250 kW,	nr	794.72	104.19		898.91
236	Single stage burner, 250 - 380 kW	nr	1306.02	138.92		1444.94
238	Single stage burner, 380 - 600 kW	nr	4497.46	173.65		4671.11
T8050	OIL BURNERS : remove existing oil burner accessory, fix only new					
640	Ignition probe (up to 30 kW)	nr		5.02		5.02
641	Ignition probe (up to 300 kW)	nr		6.69		6.69
660	Ignition sensor (up to 30 kW)	nr		5.02		5.02
661	Ignition sensor (up to 300 kW)	nr		5.02		5.02
716	Control box (15 kw to 25 kw)	nr		6.69		6.69
718	Control box (20 kw to 30 kw)	nr		6.69		6.69
724	Control box (up to 300 kw)	nr		15.06		15.06
750	Burner motor (up to 20 kW)	nr		33.47		33.47
752	Burner motor (20 kW to 30 kW)	nr		40.16		40.16
754	Burner motor (30 kW to 300 kW)	nr		60.24		60.24
756	Burner motor (over 300 kw)	nr		80.32		80.32
800	Thermocouple	nr		5.02		5.02
820	Nozzle	nr		5.02		5.02
850	Oil Filter	nr		3.35		3.35
870	Oil pump (up to 30 kW)	nr		13.39		13.39
872	Oil pump (over 300 kW)	nr		26.77		26.77
T8052	OIL BURNERS : supply only oil burner accessories					
640	Ignition probe (up to 30 kW)	nr	2.92			2.92
641	Ignition probe (up to 300 kW)	nr	7.30			7.30
660	Ignition sensor (up to 30 kW)	nr	2.92			2.92
661	Ignition sensor (up to 300 kW)	nr	7.30			7.30
716	Control box (15 kw to 25 kw)	nr	36.53			36.53
718	Control box (20 kw to 30 kw)	nr	40.91			40.91
724	Control box (up to 300 kw)	nr	219.27			219.27
750	Burner motor (up to 20 kW)	nr	40.18			40.18
752	Burner motor (20 kW to 30 kW)	nr	43.83			43.83
754	Burner motor (30 kW to 300 kW)	nr	87.66			87.66

© NSR 01 Aug 2008 - 31 Jul 2009

T34

T : MECHANICAL SERVICES - MECHANICAL HEATING/COOLING REFRIGERATION SYSTEMS : WORKS OF ALTERATION/SMALL WORKS/REPAIR			Mat. £	Lab. £	Plant £	Total £
T8052						
756	Burner motor (over 300 kw)	nr	233.74			233.74
800	Thermocouple	nr	5.20			5.20
820	Nozzle	nr	6.26			6.26
850	Oil Filter	nr	7.16			7.16
870	Oil pump (up to 30 kW)	nr	58.43			58.43
872	Oil pump (over 300 kW)	nr	211.83			211.83
T8054	OIL BURNERS AND GAS TRAINS : remove, supply and fix existing oil burner accessories					
640	Ignition probe (up to 30 kW)	nr	2.92	5.02		7.94
641	Ignition probe (up to 300 kW)	nr	7.30	6.69		13.99
660	Ignition sensor (up to 30 kW)	nr	2.92	5.02		7.94
661	Ignition sensor (up to 300 kW)	nr	7.30	5.02		12.32
716	Control box (15 kW to 25 kW)	nr	36.53	6.69		43.22
718	Control box (20 kW to 30 kW)	nr	40.91	6.69		47.60
724	Control box (up to 300 kW)	nr	219.27	15.06		234.33
750	Burner motor (up to 20 kW)	nr	40.18	33.47		73.65
752	Burner motor (20 kW to 30 kW)	nr	43.83	40.16		83.99
754	Burner motor (30 kW to 300 kW)	nr	87.66	60.24		147.90
756	Burner motor (over 300 kW)	nr	233.74	80.32		314.06
800	Thermocouple	nr	5.20	5.02		10.22
820	Nozzle	nr	6.26	5.02		11.28
850	Oil filter	nr	7.16	3.35		10.51
870	Oil pump (up to 30 kW),	nr	58.43	13.39		71.82
872	Oil pump (over 30 kW)	nr	211.83	26.77		238.60
T8056	OIL BURNERS : remove existing oil burner, fix only new					
233	Single Stage burner, 25 - 100 kw	nr		86.83		86.83
235	Single stage burner, 100 - 280 kw	nr		104.19		104.19
237	Single stage burner, 280 - 450 kw	nr		138.92		138.92
239	Single stage burner, 450 - 700 kw	nr		173.65		173.65
T8058	OIL BURNERS : supply only oil burner					
233	Single Stage burner, 25 - 100 kw	nr	468.94			468.94

© NSR 01 Aug 2008 - 31 Jul 2009

T35

T : MECHANICAL SERVICES - MECHANICAL HEATING/COOLING REFRIGERATION SYSTEMS : WORKS OF ALTERATION/SMALL WORKS/REPAIR			Mat. £	Lab. £	Plant £	Total £
T8058						
235	Single stage burner, 100 - 280 kw	nr	795.12			795.12
237	Single stage burner, 280 - 450 kw	nr	1306.02			1306.02
239	Single stage burner, 450 - 700 kw	nr	2147.46			2147.46
T8060	**OIL BURNERS : remove existing oil burner, supply and fix new**					
233	Single stage burner, 25 - 100 kW	nr	468.94	86.83		555.77
235	Single stage burner, 100 - 280 kW	nr	795.12	104.19		899.31
237	Single stage burner, 280 - 450 kW	nr	1306.02	138.92		1444.94
239	Single stage burner, 450 - 700 kW	nr	2147.46	173.65		2321.11
T8062	**FIRE VALVE ACCESSORIES**					
500	Remove existing, fix only new fire valve linkage and fusible link assembly	nr		20.08		20.08
505	Supply only fire valve linkage and fusible link assembly	nr	517.43			517.43
T8064	**FIRE VALVE ACCESSORIES : remove existing, supply and fix new**					
500	Fire valve linkage and fusible link assembly	nr	499.79	20.08		519.87
T8066	**FIRE VALVE ACCESSORIES : remove damaged pressure / thermometer / contents gauge, fix only new**					
520	Dial type	nr		30.12		30.12
522	Stem type	nr		30.12		30.12
T8068	**FIRE VALVE ACCESSORIES : supply only pressure / thermometer / contents gauge**					
520	Dial type	nr	138.60			138.60
522	Stem type	nr	41.15			41.15
T8070	**FIRE VALVE ACCESSORIES : remove damaged dial type, pressure/thermometer/contents gauge, supply and fix new**					
520	Dial type	nr	138.60	30.12		168.72
522	Stem type	nr	41.15	30.12		71.27
T8072	**FIRE VALVE ACCESSORIES : remove damaged automatic air vent, fix only new**					
530	15 mm Diameter	nr		70.28		70.28

T : MECHANICAL SERVICES - MECHANICAL HEATING/COOLING REFRIGERATION SYSTEMS : WORKS OF ALTERATION/SMALL WORKS/REPAIR			Mat. £	Lab. £	Plant £	Total £
T8074	FIRE VALVE ACCESSORIES : supply only automatic air vent					
530	15 mm Diameter	nr	123.12			123.12
T8076	FIRE VALVE ACCESORIES : remove damaged, automatic air vent, supply and fix new					
530	15 mm Diameter	nr	123.12	70.28		193.40
T8078	VALVES AND ANCILLARIES : remove damaged safety valve, fix only new					
116	15 - 25 mm Diameter	nr		44.18		44.18
118	32 - 50 mm Diameter	nr		74.30		74.30
120	65 mm Diameter	nr		84.34		84.34
122	80 mm Diameter	nr		106.42		106.42
124	100 mm Diameter	nr		134.20		134.20
T8080	VALVES AND ANCILLARIES : supply only safety valve					
116	15 - 25 mm Diameter	nr	226.84			226.84
118	32 - 50 mm Diameter	nr	550.09			550.09
120	65 mm Diameter	nr	619.32			619.32
122	80 mm Diameter	nr	902.27			902.27
T8082	VALVES AND ANCILLIARIES : remove damaged safety valve, supply and fix new					
116	15 - 25 mm Diameter	nr	226.84	44.18		271.02
118	32 - 50 mm Diameter	nr	550.09	74.30		624.39
120	65 mm Diameter	nr	619.32	84.34		703.66
122	80 mm Diameter	nr	902.27	106.42		1008.69
T8084	VALVES AND ANCILLARIES : remove damaged three way vent cock, fix only new					
116	15 - 25 mm Diameter	nr		31.26		31.26
118	32 - 50 mm Diameter	nr		43.41		43.41
120	65 mm Diameter	nr		55.57		55.57
T8086	VALVES AND ANCILLARIES : supply only three way vent cock					
116	15 - 25 mm Diameter	nr	297.12			297.12
118	32 - 50 mm Diameter	nr	717.44			717.44
120	65 mm Diameter	nr	907.81			907.81

© NSR 01 Aug 2008 - 31 Jul 2009

T37

T : MECHANICAL SERVICES - MECHANICAL HEATING/COOLING REFRIGERATION SYSTEMS : WORKS OF ALTERATION/SMALL WORKS/REPAIR			Mat. £	Lab. £	Plant £	Total £
T8088	VALVES AND ANCILLIARIES : remove damaged three way vent cock, supply and fix new					
116	15 - 25 mm diameter	nr	297.12	31.26		328.38
118	32 - 50 mm diameter	nr	717.44	43.41		760.85
120	65 mm diameter	nr	907.81	55.57		963.38

			Mat. £	Lab. £	Plant £	Total £
	U : MECHANICAL SERVICES - VENTILATION/AIR CONDITIONING SYSTEMS : WORKS OF ALTERATION/SMALL WORKS/REPAIR					
	U32 : VENTILATION DUCTWORK					
U3200	**STRAIGHT DUCTWORK : supply and fit rectangular section, low velocity, 0.6 mm thick, stiffners, joints, couplers and supports included**					
100	Sum of two sides 200 mm	nr	28.83	10.77		39.60
102	Sum of two sides 250 mm	nr	30.48	10.77		41.25
103	Sum of two sides 300 mm	nr	32.11	10.77		42.88
104	Sum of two sides 350 mm	nr	33.21	10.77		43.98
105	Sum of two sides 400 mm	nr	34.28	11.31		45.59
106	Sum of two sides 450 mm	nr	35.39	11.31		46.70
107	Sum of two sides 500 mm	nr	33.86	11.85		45.71
108	Sum of two sides 550 mm	nr	37.55	11.85		49.40
109	Sum of two sides 600 mm	nr	38.09	12.39		50.48
110	Sum of two sides 650 mm	nr	39.20	12.92		52.12
111	Sum of two sides 700 mm	nr	40.83	13.46		54.29
112	Sum of two sides 750 mm	nr	42.46	14.00		56.46
113	Sum of two sides 800 mm	nr	43.00	14.54		57.54
U3210	**STRAIGHT DUCTWORK : supply and fit circular section, low velocity, 0.6mm thick, stiffners, joints, couplers and supports included**					
200	160mm diameter	m	13.01	3.64		16.65
202	180mm diameter	m	15.71	3.84		19.55
204	200mm diameter	m	16.57	3.84		20.41
206	250mm diameter	m	19.34	4.24		23.58
208	315mm diameter	m	22.88	4.65		27.53
210	350mm diameter	m	25.72	6.26		31.98
212	400mm diameter	m	31.32	6.46		37.78
214	450mm diameter	m	34.51	6.87		41.38
216	500mm diameter	m	37.47	7.27		44.74
U3220	**DUCTWORK FITTINGS : supply and fit bend, rectangular section, low velocity, 0.6 mm thick**					
100	Sum of two sides 200 mm	nr	45.06	8.08		53.14

© NSR 01 Aug 2008 - 31 Jul 2009   U1

U : MECHANICAL SERVICES - VENTILATION/AIR CONDITIONING SYSTEMS : WORKS OF ALTERATION/SMALL WORKS/REPAIR			Mat. £	Lab. £	Plant £	Total £
U3220						
102	Sum of two sides 250 mm	nr	50.03	8.08		58.11
103	Sum of two sides 300 mm	nr	50.26	8.08		58.34
104	Sum of two sides 350 mm	nr	50.52	8.08		58.60
105	Sum of two sides 400 mm	nr	50.75	8.62		59.37
106	Sum of two sides 450 mm	nr	52.80	8.62		61.42
107	Sum of two sides 500 mm	nr	53.89	9.15		63.04
108	Sum of two sides 550 mm	nr	54.43	9.15		63.58
109	Sum of two sides 600 mm	nr	54.98	9.69		64.67
110	Sum of two sides 650 mm	nr	56.07	10.23		66.30
111	Sum of two sides 700 mm	nr	58.68	10.77		69.45
112	Sum of two sides 750 mm	nr	58.96	11.31		70.27
113	Sum of two sides 800 mm	nr	65.36	11.85		77.21
U3230	DUCTWORK FITTINGS : supply and fit branch, rectangular section, low velocity, 0.6 mm thick					
100	Sum of two sides 200 mm	nr	42.46	4.31		46.77
102	Sum of two sides 250 mm	nr	42.65	4.31		46.96
103	Sum of two sides 300 mm	nr	42.86	4.31		47.17
104	Sum of two sides 350 mm	nr	42.99	4.31		47.30
105	Sum of two sides 400 mm	nr	43.20	4.85		48.05
106	Sum of two sides 450 mm	nr	43.41	4.85		48.26
107	Sum of two sides 500 mm	nr	45.17	4.85		50.02
108	Sum of two sides 550 mm	nr	45.39	5.38		50.77
109	Sum of two sides 600 mm	nr	45.60	5.38		50.98
110	Sum of two sides 650 mm	nr	45.73	5.38		51.11
111	Sum of two sides 700 mm	nr	45.93	5.92		51.85
112	Sum of two sides 750 mm	nr	46.16	5.92		52.08
113	Sum of two sides 800 mm	nr	46.26	5.92		52.18
U3240	DUCTWORK FITTINGS : supply and fit bend, circular section, low velocity, 0.6mm thick					
300	160mm Diameter	nr	21.01	1.41		22.42
302	180mm diameter	nr	32.32	1.62		33.94
304	200mm diameter	nr	23.82	1.62		25.44

© NSR 01 Aug 2008 - 31 Jul 2009   U2

U : MECHANICAL SERVICES - VENTILATION/AIR CONDITIONING SYSTEMS : WORKS OF ALTERATION/SMALL WORKS/REPAIR			Mat. £	Lab. £	Plant £	Total £
U3240						
306	250mm diameter	nr	31.40	2.42		33.82
308	300mm diameter	nr	38.98	2.63		41.61
310	350mm diameter	nr	43.31	3.43		46.74
312	400mm diameter	nr	61.72	3.84		65.56
314	450mm diameter	nr	72.55	3.84		76.39
316	500mm diameter	nr	83.37	4.44		87.81
U3250	DUCTWORK FITTINGS : supply and fit branch, circular section, low velocity, 0.6mm thick					
300	160mm Diameter	nr	34.25	2.83		37.08
302	180mm diameter	nr	34.25	3.84		38.09
304	200mm diameter	nr	46.83	4.24		51.07
306	250mm diameter	nr	46.83	5.45		52.28
308	300mm diameter	nr	64.26	6.06		70.32
310	350mm diameter	nr	64.26	8.08		72.34
312	400mm diameter	nr	84.64	8.28		92.92
314	450mm diameter	nr	84.64	8.28		92.92
316	500mm diameter	nr	84.64	9.29		93.93
U3260	GRILLES AND DIFFUSERS : install only extract grille, aluminium finish, adjustable blades, no volume control, fitting to ductwork					
100	Sum of two sides 200 mm	nr		8.23		8.23
102	Sum of two sides 250 mm	nr		8.23		8.23
103	Sum of two sides 300 mm	nr		8.23		8.23
104	Sum of two sides 350 mm	nr		8.23		8.23
105	Sum of two sides 400 mm	nr		8.78		8.78
106	Sum of two sides 450 mm	nr		8.78		8.78
107	Sum of two sides 500 mm	nr		9.33		9.33
108	Sum of two sides 550 mm	nr		9.33		9.33
109	Sum of two sides 600 mm	nr		9.88		9.88
110	Sum of two sides 650 mm	nr		10.42		10.42
111	Sum of two sides 700 mm	nr		10.97		10.97
112	Sum of two sides 750 mm	nr		11.52		11.52
113	Sum of two sides 800 mm	nr		12.07		12.07

© NSR 01 Aug 2008 - 31 Jul 2009

U3

U : MECHANICAL SERVICES - VENTILATION/AIR CONDITIONING SYSTEMS : WORKS OF ALTERATION/SMALL WORKS/REPAIR			Mat. £	Lab. £	Plant £	Total £
	U45 : AIR CONDITIONING					
U4500	**FAN COIL UNITS : fix fan coil unit, 665 mm deep x 235 mm high**					
100	710 mm long, 3 row cooling only	nr		247.65		247.65
102	910 mm long, 3 row cooling only	nr		267.73		267.73
104	1280 mm long, 3 row cooling only	nr		287.81		287.81
106	1460 mm long, 3 row cooling only	nr		307.89		307.89
108	1900 mm long, 3 row cooling only	nr		327.97		327.97
110	710 mm long, 4 row cooling only	nr		254.35		254.35
112	910 mm long, 4 row cooling only	nr		274.43		274.43
114	1280 mm long, 4 row cooling only	nr		294.51		294.51
116	1460 mm long, 4 row cooling only	nr		314.59		314.59
118	1900 mm long, 4 row cooling only	nr		334.67		334.67
124	710 mm long, 3 row cooling, 1 row heating	nr		257.69		257.69
126	910 mm long, 3 row cooling, 1 row heating	nr		274.43		274.43
128	1280 mm long, 3 row cooling, 1 row heating	nr		294.51		294.51
130	1460 mm long, 3 row cooling, 1 row heating	nr		314.59		314.59
132	1900 mm long, 3 row cooling, 1 row heating	nr		334.67		334.67
134	710 mm long, 4 row cooling, 1 row heating	nr		264.39		264.39
136	910 mm long, 4 row cooling, 1 row heating	nr		281.12		281.12
138	1280 mm long, 4 row cooling, 1 row heating	nr		301.20		301.20
140	1460 mm long, 4 row cooling, 1 row heating	nr		321.28		321.28
142	1900 mm long, 4 row cooling, 1 row heating	nr		341.36		341.36
U4510	**FAN COIL UNITS : supply fan coil unit, 665 mm deep x 235 mm high**					
100	710 mm long, 3 row cooling only	nr	622.00			622.00
102	910 mm long, 3 row cooling only	nr	686.00			686.00
104	1280 mm long, 3 row cooling only	nr	756.00			756.00
106	1460 mm long, 3 row cooling only	nr	908.00			908.00
108	19000 mm long, 3 row cooling only	nr	1024.00			1024.00
110	710 mm long, 4 row cooling only	nr	666.00			666.00
112	910 mm long, 4 row cooling only	nr	732.00			732.00
114	1280 mm long, 4 row cooling only	nr	808.00			808.00
116	1460 mm long, 4 row cooling only	nr	958.00			958.00
118	1900 mm long, 4 row cooling only	nr	1092.00			1092.00

© NSR 01 Aug 2008 - 31 Jul 2009   U4

U : MECHANICAL SERVICES - VENTILATION/AIR CONDITIONING SYSTEMS : WORKS OF ALTERATION/SMALL WORKS/REPAIR			Mat. £	Lab. £	Plant £	Total £
U4510						
124	710 mm long, 3 row cooling, 1 row heating	nr	622.00			622.00
126	910 mm long, 3 row cooling, 1 row heating	nr	686.00			686.00
128	1280 mm long, 3 row cooling, 1 row heating	nr	756.00			756.00
130	1460 mm long, 3 row cooling, 1 row heating	nr	908.00			908.00
132	1900 mm long, 3 row cooling, 1 row heating	nr	1024.00			1024.00
134	710 mm long, 4 row cooling, 1 row heating	nr	666.00			666.00
136	910 mm long, 4 row cooling, 1 row heating	nr	732.00			732.00
138	1280 mm long, 4 row cooling, 1 row heating	nr	958.00			958.00
140	1460 mm long, 4 row cooling, 1 row heating	nr	958.00			958.00
142	1900 mm long, 4 row cooling, 1 row heating	nr	1092.00			1092.00
U4520	FAN COIL UNITS : supply and fix fan coil unit, 665 mm deep x 235 mm high					
100	710 mm long, 3 row cooling only	nr	622.00	247.65		869.65
102	910 mm long, 3 row cooling only	nr	686.00	267.73		953.73
104	1280 mm long, 3 row cooling only	nr	756.00	287.81		1043.81
106	1460 mm long, 3 row cooling only	nr	908.00	307.89		1215.89
108	19000 mm long, 3 row cooling only	nr	1024.00	327.97		1351.97
110	710 mm long, 4 row cooling only	nr	666.00	254.35		920.35
112	910 mm long, 4 row cooling only	nr	732.00	274.43		1006.43
114	1280 mm long, 4 row cooling only	nr	808.00	294.51		1102.51
116	1460 mm long, 4 row cooling only	nr	958.00	314.59		1272.59
118	1900 mm long, 4 row cooling only	nr	1092.00	334.67		1426.67
124	710 mm long, 3 row cooling, 1 row heating	nr	622.00	257.69		879.69
126	910 mm long, 3 row cooling, 1 row heating	nr	686.00	274.43		960.43
128	1280 mm long, 3 row cooling, 1 row heating	nr	756.00	294.51		1050.51
130	1460 mm long, 3 row cooling, 1 row heating	nr	908.00	314.59		1222.59
132	1900 mm long, 3 row cooling, 1 row heating	nr	1024.00	334.67		1358.67
134	710 mm long, 4 row cooling, 1 row heating	nr	666.00	264.39		930.39
136	910 mm long, 4 row cooling, 1 row heating	nr	732.00	281.12		1013.12
138	1280 mm long, 4 row cooling, 1 row heating	nr	958.00	301.20		1259.20
140	1460 mm long, 4 row cooling, 1 row heating	nr	958.00	321.28		1279.28
142	1900 mm long, 4 row cooling, 1 row heating	nr	1092.00	341.36		1433.36

© NSR 01 Aug 2008 - 31 Jul 2009

U5

U : MECHANICAL SERVICES - VENTILATION/AIR CONDITIONING SYSTEMS : WORKS OF ALTERATION/SMALL WORKS/REPAIR			Mat. £	Lab. £	Plant £	Total £
U4566	A/C UNITS : Supply and fix cassette type unit, cooling only, complete with remote control external condensers, fixings and brackets, pre-charged refrigerant (up to 10m) DAIKIN Refs					
701	3.4 kW Ref:FCQ35C/RKS35E	nr	1147.00	204.15		1351.15
702	5.0 kW Ref: FCQ50C/RKS50F	nr	1197.00	204.15		1401.15
704	5.7 kW Ref: FCQ60C/RKS60F	nr	1277.00	204.15		1481.15
U4568	A/C UNITS : Fix only cassette type unit, cooling only, complete with remote control, external condensers, fixings and brackets, pre-charged refrigerant (up to 10m)					
701	3.4 kW	nr		204.15		204.15
702	5.0 kW	nr		204.15		204.15
704	5.7 kW	nr		204.15		204.15
U4572	A/C UNITS : Supply and fix cassette type unit, heat pump, complete with remote control, external condensing unit, fixings and brackets, pre-charged refrigerant (up to 10m) DAIKIN Refs					
701	3.4 kW (Ref: FCQ35C/RXS35E)	nr	1127.00	204.15		1331.15
702	5.7 kW (Ref:FCQ60C/RXS60F)	nr	1337.00	204.15		1541.15
704	10 kW (Ref:FCQ100C/RZQ100BV3)	nr	1889.00	204.15		2093.15
U4574	A/C UNITS : fix only cassette type unit, heat pump, complete with remote control, external condensers, fixings and brackets, pre-charged refrigerant (up to 10m)					
700	3.4 kW	nr		204.15		204.15
720	5.7 kW	nr		204.15		204.15
722	10 kW	nr		204.15		204.15
U4578	A/C UNITS : supply and fix high wall type unit, cooling only, complete with remote control, external condensers, fixings and brackets, pre-charged refrigerant (up to 10m) DAIKIN Refs					
800	3.4 kW FTKS35DW/RKS35E	nr	623.00	204.15		827.15
810	6.0 kW FTKS60F/RKS60F	nr	858.00	204.15		1062.15
812	7.1 kW FTKS71F/RKS71F	nr	1153.00	204.15		1357.15
U4580	A/C UNITS : fix only high wall type unit, cooling only, complete with remote control, external condensers, fixings and brackets, pre-charged refrigerant (up to 10m)					
800	3.4 kW	nr		204.15		204.15
810	6.0 kW	nr		204.15		204.15
812	7.1 kW	nr		204.15		204.15

© NSR 01 Aug 2008 - 31 Jul 2009   U6

	U : MECHANICAL SERVICES - VENTILATION/AIR CONDITIONING SYSTEMS : WORKS OF ALTERATION/SMALL WORKS/REPAIR		Mat. £	Lab. £	Plant £	Total £
U4584	A/C UNITS : Supply and fix high wall type unit, complete with remote control, external condensing unit, fixings and brackets, pre-charged refrigerant (up to 10m) DAIKIN Refs					
820	3.40 kW Ref: FTXS35DL/RXS35E	nr	678.00	204.15		882.15
826	6.0 kW (Ref: FTXS60F/RXS60F)	nr	938.00	204.15		1142.15
828	7.1 kW (Ref: FTXS71F/RXS71F)	nr	1233.00	204.15		1437.15
830	10 kW (Ref: FAQ100B/RZQ100BV3)	nr	1889.00	204.15		2093.15
U4586	A/C UNITS : fix only high wall type unit, heat pump, complete with remote control, external condensers, fixings and brackets, pre-charged refrigerant (up to 10m)					
820	3.4 kW	nr		204.15		204.15
826	6.0kW	nr		204.15		204.15
828	7.1 kW	nr		204.15		204.15
830	10 kW	nr		204.15		204.15
U4590	CONDENSATE PUMPS: Remove existing condensate pump, fix only new.					
010	Auto Condensate pump, compact tank, maxflow 66GPH	nr		20.08		20.08
012	Auto Condensate pump, medium tank, maxflow 66GPH	nr		20.08		20.08
014	Auto Condensate pump, large tank, maxflow 212GPH	nr		23.43		23.43
016	Auto Condensate pump, large tank, maxflow 357GPH	nr		23.43		23.43
020	Universal Peristaltic pump, 240V	nr		20.08		20.08
U4592	CONDENSATE PUMPS: Supply only condensate pump.					
010	Auto Condensate pump, compact tank, maxflow 66GPH	nr	113.00			113.00
012	Auto Condensate pump, medium tank, maxflow 66GPH	nr	128.00			128.00
014	Auto Condensate pump, large tank, maxflow 212GPH	nr	137.00			137.00
016	Auto Condensate pump, large tank, maxflow 357GPH	nr	293.00			293.00
020	Universal Peristaltic pump, 240V	nr	93.46			93.46
U4594	CONDENSATE PUMPS: Remove existing condensate pump, supply and fix new.					
010	Auto Condensate pump, compact tank, maxflow 66GPH	nr	113.00	20.08		133.08
012	Auto Condensate pump, medium tank, maxflow 66GPH	nr	128.00	20.08		148.08
014	Auto Condensate pump, large tank, maxflow 212GPH	nr	137.00	23.43		160.43
016	Auto Condensate pump, large tank, maxflow 357GPH	nr	293.00	23.43		316.43
020	Universal Peristaltic pump, 240V	nr	93.46	20.08		113.54

© NSR 01 Aug 2008 - 31 Jul 2009

U7

U : MECHANICAL SERVICES - VENTILATION/AIR CONDITIONING SYSTEMS : WORKS OF ALTERATION/SMALL WORKS/REPAIR			Mat. £	Lab. £	Plant £	Total £
	U55 : AIR CONDITIONING REPAIRS					
U5524	REFRIGERANT GAS : remove refrigerant from circuit, re-charge re-using refrigerant, up to 20 metres between A/C unit and condensers					
105	R22	nr		204.15		204.15
110	R-410A	nr		204.15		204.15
U5530	REFRIGERANT GAS : supply and install additional refrigerant gas to air conditioning unit					
150	Type R22	kg	21.00	28.45		49.45
160	Type R-410A	kg	30.00	28.45		58.45
U5536	UNIT PARTS : supply only item of equipment					
150	Compressor, up to 3kW	nr	355.00			355.00
160	Compressor, 3 - 6 kW	nr	680.00			680.00
170	Compressor, over 6 kW	nr	950.00			950.00
180	Sump Pump, 2ltr	nr	95.00			95.00
250	Fan motor, 1 kw	nr	96.64			96.64
270	Impeller	nr	18.36			18.36
290	Fan Blade	nr	18.36			18.36
U5542	UNIT PARTS : supply and fix item of equipment, cassette or high wall type A/C unit.					
150	Compressor, up to 3kW	nr	355.00	140.56		495.56
160	Compressor, 3 - 6 kW	nr	680.00	140.56		820.56
170	Compressor, over 6 kW	nr	950.00	140.56		1090.56
180	Sump Pump, 2ltr	nr	95.00	18.41		113.41
250	Fan motor, 1 kwt.	nr	96.64	69.12		165.76
270	Impeller	nr	18.36	51.84		70.20
290	Fan Blade	nr	18.36	17.28		35.64
U5544	UNIT PARTS : remove faulty or damaged item of equipment, install only new item of equipment, cassette or high wall type A/C unit					
150	Compressor, up to 3 kW	nr		140.56		140.56
160	Compressor, 3 - 6 kW	nr		140.56		140.56
170	Compressor, over 6 kW	nr		140.56		140.56
180	Sump Pump, 2ltr	nr		18.41		18.41

© NSR 01 Aug 2008 - 31 Jul 2009   U8

U : MECHANICAL SERVICES - VENTILATION/AIR CONDITIONING SYSTEMS : WORKS OF ALTERATION/SMALL WORKS/REPAIR			Mat. £	Lab. £	Plant £	Total £
U5544						
190	PCB	nr		34.56		34.56
250	Fan motor, 1 kw	nr		69.12		69.12
270	Impeller	nr		51.84		51.84
290	Fan blade	nr		17.28		17.28
300	Compressor, evaporator, 4 way valve and TX valve assembly	nr		103.68		103.68
460	Expansion valve	nr		120.96		120.96
U5546	**COMPRESSORS**					
110	Pressure test and evacuation	nr	26.00	125.50		151.50
120	Re-charge and re-commission compressor	nr		60.24		60.24

© NSR 01 Aug 2008 - 31 Jul 2009

U9

U : MECHANICAL SERVICES - VENTILATION/AIR CONDITIONING SYSTEMS : WORKS OF ALTERATION/SMALL WORKS/REPAIR	Mat. £	Lab. £	Plant £	Total £
U : MECHANICAL SERVICES - VENTILATION/AIR CONDITIONING SYSTEMS : WORKS OF ALTERATION/SMALL WORKS/REPAIR				U10

MECHANICAL

BASIC PRICES: LABOUR, PLANT AND MATERIALS

BASIC PRICES : LABOUR, PLANT AND MATERIALS

LABOUR

Kango Operator	KG	Hr	£14.81
Bricklayer	BR	Hr	£17.28
Bricklayers Labourer	BL	Hr	£12.93
Scaffolder	SC	Hr	£17.28
Painter	PA	Hr	£17.28
General Labourer	GL	Hr	£12.17
Heating Engineer	HE	Hr	£20.08
Heating Engineer Apprentice	HEA	Hr	£14.65
Advanced Plumber	AP	Hr	£24.27
Trained plumber (Mechanical & Electrical)	TP	Hr	£20.08
Apprentice plumber	PPA	Hr	£14.65
Thermal insulation engineer	IE	Hr	£15.85
Duct fitter	DF	Hr	£16.46
Fitters mate	FM	Hr	£15.85
Electrical (Mechanical & Electrical)	E	Hr	£25.38
Electrical Apprentice	EA	Hr	£20.15
Electrical Labourer	EM	Hr	£19.08

There has been a rationalisation of rates e.g. Heating Engineer is the same rate as Trained Plumber, Electrical (Mechanical and Electrical) is the same rate as Electrician, etc., etc.

BASIC PRICES : LABOUR, PLANT AND MATERIALS

A: MECHANICAL SERVICES - CONTRACTORS GENERAL COST ITEMS : WORKS OF ALTERATION/SMALL WORKS/REPAIR

GENERALLY
Generally

Generally; Electric radiator	Wk	19.00
Generally; Festoon lighting	Wk	32.00
Generally; Scaffold alarm, Quad beam & passive, 50m range - Hire per week	Wk	21.00
Generally; Scaffold alarm, Quad beam & passive, 50m range- set up and 4 weeks hire	Item	234.00
Generally; Scaffold lighting - Security flood lighting - per light	Wk	3.50
Generally; Scaffold lighting - Security flood lighting - set up & 4 weeks hire	nr	50.00
Generally; Genie super lift 363kg	Day	102.00

LIGHTWEIGHT ALUMIMIUM ACCESS UNITS
Lightweight alumimium access units

Lightweight alumimium access units; Chimney scaffold to full surround of centre ridge stack	Wk	220.00
Lightweight alumimium access units; Chimney scaffold to full surround of centre ridge stack	Day	110.00
Lightweight alumimium access units; Chimney scaffold to Half of centre ridge stack	Wk	110.00
Lightweight alumimium access units; Chimney scaffold to half of centre ridge stack	Day	55.00
Lightweight alumimium access units; Compact scissor lift 7.8m	Day	216.00
Lightweight alumimium access units; Compact scissor lift 7.8m	Wk	360.00
Lightweight alumimium access units; Staircase access unit with 300 - 450mm wide platform	Wk	90.00
Lightweight alumimium access units; Staircase access unit with 300 - 450mm wide platform	Day	45.00
Lightweight alumimium access units; Staircase access unit with 600 - 675mm wide platform	Wk	90.00
Lightweight alumimium access units; Staircase access unit with 600 - 675mm wide platform	Day	45.00
Lightweight alumimium access units; Window scaffold with 450mm wide platform	Wk	90.00

© NSR 2008-2009

BASIC PRICES : LABOUR, PLANT AND MATERIALS

Lightweight alumimium access units; Window scaffold with 450mm wide platform	Day	45.00
Lightweight alumimium access units; Window scaffold with 600mm wide platform	Wk	90.00
Lightweight alumimium access units; Window scaffold with 600mm wide platform	Day	45.00

NON - MECHANICAL PLANT
Non - Mechanical Plant

Non - Mechanical Plant; Adjustable base plates per 10 No	Hr	0.13
Non - Mechanical Plant; Brick guards per 10 No	Hr	0.16
Non - Mechanical Plant; Independent tied scaffold complete (per m^2)	Wk	3.58
Non - Mechanical Plant; Putlog scaffold complete (per m^2)	Wk	2.71
Non - Mechanical Plant; Scaffold boards per 100 feet	Hr	0.55
Non - Mechanical Plant; Scaffold fittings per 10 No	Wk	5.22
Non - Mechanical Plant; Scaffold tubes per 100 feet	Hr	0.40
Non - Mechanical Plant; Security fencing 3.5m x 2.2m panels, concrete feet	Hr	0.18
Non - Mechanical Plant; Tarpaulin 5 x 4m £17.00/wk (rate below per m2)	Wk	0.90

SCAFFOLD TOWER ON CASTORS 2.50 X 0.85M ON PLAN
Scaffold tower on castors 2.50 x 0.85m on plan

Scaffold tower on castors 2.50 x 0.85m on plan; 3.20m total Height	Day	56.75
Scaffold tower on castors 2.50 x 0.85m on plan; 3.20m total Height	Wk	113.50
Scaffold tower on castors 2.50 x 0.85m on plan; 4.20m total Height	Day	66.50
Scaffold tower on castors 2.50 x 0.85m on plan; 4.20m total Height	Wk	133.00
Scaffold tower on castors 2.50 x 0.85m on plan; 5.20m total Height	Day	76.25
Scaffold tower on castors 2.50 x 0.85m on plan; 5.20m total Height	Wk	152.50

BASIC PRICES : LABOUR, PLANT AND MATERIALS

Scaffold tower on castors 2.50 x 0.85m on plan; 6.20m total Height	Wk	172.00
Scaffold tower on castors 2.50 x 0.85m on plan; 6.20m total height	Day	86.00

SCAFFOLD TOWER ON CASTORS 2.50 X 1.45M ON PLAN
Scaffold tower on castors 2.50 x 1.45m on plan

Scaffold tower on castors 2.50 x 1.45m on plan; 3.20m total Height	Wk	113.50
Scaffold tower on castors 2.50 x 1.45m on plan; 3.20m total Height	Day	56.75
Scaffold tower on castors 2.50 x 1.45m on plan; 4.20m total Height	Wk	133.00
Scaffold tower on castors 2.50 x 1.45m on plan; 4.20m total Height	Day	66.50
Scaffold tower on castors 2.50 x 1.45m on plan; 5.20m total Height	Day	76.25
Scaffold tower on castors 2.50 x 1.45m on plan; 5.20m total Height	Wk	152.50
Scaffold tower on castors 2.50 x 1.45m on plan; 6.20m total Height	Wk	172.00
Scaffold tower on castors 2.50 x 1.45m on plan; 6.20m total height	Day	86.00

P:BUILDING FABRIC SUNDRIES

BRICKWORK AND BLOCKWORK
Brickwork and Blockwork

Brickwork and Blockwork; 5KVA Diesel Generator	Min	0.06
Brickwork and Blockwork; Dry Diamond Driller	Min	0.17
Brickwork and Blockwork; Kango hammer type 2500	Min	0.02
Brickwork and Blockwork; Sundry bits, cutters etc	Item	1.00

CONCRETE WORK
Concrete Work

Concrete Work; Disc cutter/angle grinder	Min	0.02
Concrete Work; Metal disc	nr	13.61

© NSR 2008-2009

BASIC PRICES : LABOUR, PLANT AND MATERIALS

R:PLUMBING : WORKS OF ALTERATION/SMALL WORKS/REPAIR
RW & SOIL PLASTICS

GEBERIT RAINWATER

<NONE>; 110MM 2110.4 GREY SHOE	EACH	11.68
<NONE>; 110MM 2151.4 GREY JOINT BKT GUTTER FITTING	EACH	2.40
<NONE>; 110MM 2152.4 GREY SUPPORT BKT GUTTER FITTING	EACH	0.97
<NONE>; 110MM 2153.4.25 GREY RUNNING OUTLET GUTTER FITTING	EACH	3.54
<NONE>; 110MM 2155.4 GREY SHORT STOP END	EACH	1.77
<NONE>; 131MM 2451.5 BROWN JOINT BRACKET GUTTER FITTING	EACH	3.34
<NONE>; 131MM 2452.5 BROWN SUPPORT BRACKET GUTTER FITTING	EACH	1.42
<NONE>; 131MM 2453.5.25 BROWN RUNNING OUTLET GUTTER FITTING	EACH	5.31
<NONE>; 131MM 2455.5 BROWN SHORT STOP END	EACH	2.72
<NONE>; 131MM BROWN 4M GUTTER RAPIDFLOW RAINWATER SYSTEM 2450.5.40B	EACH	17.70
<NONE>; 4M 110MM 2150.4.40 GREY GUTTER TRUE HALF ROUND	EACH	13.61
<NONE>; 4M 62MM 2200.23.40 BLACK SQUARE DOWNPIPE	EACH	15.32
<NONE>; 4M 68MM 2100.25.40 GREY ROUND DOWNPIPE	EACH	15.18
<NONE>; 62MM 2210.23 BLACK SHOE	EACH	3.12
<NONE>; 62MM 2213.23 BLACK ADJUSTABLE PIPE FITTING CLIP - DOWNPIPE	EACH	1.29
<NONE>; 92DEG 63MM 2208.23.92 BKL BEND	EACH	3.39
<NONE>; 92DEG 68MM 2108.25.92 GREY BEND	EACH	3.01

GEBERIT SOIL & WASTE

<NONE>; 110MM 100P.4 GREY CIRCULAR DPPE 4 METRES	EACH	41.22
<NONE>; 110MM 101P.4 GREY 92.5 DEG DPPE BEND - DOWNPIPE FITTINGS	EACH	16.11
<NONE>; 110MM 110.4 GREY STR COUPLING 2XLIQUID WELD SOCKETS	EACH	7.69
<NONE>; 110MM 143.4 GREY PIPE BRACKET	EACH	3.34
<NONE>; 110MMX4M 100.4.40 GREY SOIL PIPE	EACH	47.24
<NONE>; 160MM 110.6 GEY STR COUPLING 2XLIQUID WELD SOCKETS	EACH	22.17
<NONE>; 160MMX4M 100.6.40 GREY SOIL PIPE	EACH	122.49
<NONE>; 32MM 210.125 BLACK STR CPLNG 2 X LIQUID WELD SOCKET	EACH	1.60
<NONE>; 32MM 225.125 GREY EXP COUPLING LIQUID WELD AND SEAL RING SKT	EACH	2.81

BASIC PRICES : LABOUR, PLANT AND MATERIALS

R:PLUMBING : WORKS OF ALTERATION/SMALL WORKS/REPAIR
RW & SOIL PLASTICS

GEBERIT SOIL & WASTE

<NONE>; 32MM 611.125 WHITE BOTTLE "P" TRAP 75MM SEAL TRAP	EACH	8.15
<NONE>; 32MM 632.125 WHITE "S" TRAP 75MM SEAL TRAP	EACH	9.27
<NONE>; 40MM 210.15 GREY STR COUPLING 2 X LIQUID WELD SOCKET	EACH	1.59
<NONE>; 40MM 225.15 GREY EXP COUPLING LIQUID WELD AND SEAL RING SKT	EACH	3.39
<NONE>; 40MM 611.15 WHITE BOTTLE "P" TRAP 75MM SEAL TRAP	EACH	9.72
<NONE>; 40MM 632.15 WHITE "S" TRAP 75MM SEAL TRAP	EACH	10.84
<NONE>; 4M 32MM 200.125.40 GREY WASTE PIPE	EACH	10.23
<NONE>; 4M 40MM 200.15.40 GREY WASTE PIPE	EACH	12.66
<NONE>; 4M 50MM 200.2.40 GREY WASTE PIPE	EACH	19.09
<NONE>; 50MM 152.039.16.1 WHITE "P" TRAP	EACH	17.64
<NONE>; 50MM 210.2 GREY STR COUPLING 2 X LIQUID WELD SOCKET	EACH	2.91
<NONE>; 50MM 225.2 GREY EXP COUPLING LIQUID WELD AND SEAL RING SKT	EACH	4.59
<NONE>; 82MM 110.3 GREY STR COUPLING 2XLIQUID WELD SOCKETS	EACH	6.15
<NONE>; 82MMX4M 100.3.40 GREY SOIL PIPE	EACH	46.38
<NONE>; 90MM 366.730.16.1 WHITE "P" TRAP	EACH	52.00

HEPWORTH RAINWATER

<NONE>; 112MM BLACK 2M SQUARE GUTTER BRACKET SG13/3BL	EACH	2.17
<NONE>; 112MM BLACK 2M SQUARE GUTTER RUNNING OUTLET SG16BL	EACH	8.81
<NONE>; 112MM BLACK 2M SQUARE GUTTER SG2BL	EACH	13.16
<NONE>; 112MM BLACK 2M SQUARE GUTTER STOP END SG17BL	EACH	3.06
<NONE>; 112MM BLACK 2M SQUARE GUTTER UNION SG12BL	EACH	5.73

MARLEY: RAINWATER

<NONE>; 150MM REW2G GREY STOP END GUTTER - HALF ROUND	EACH	5.25
<NONE>; 150MM RGW4G GREY 4M GUTTER HALF ROUND RAINWATER SYSTEM	PER METRE	31.50
<NONE>; 150MM RKW1G GREY FASCIA BKT HALF ROUND RAINWATER SYSTEM	EACH	2.58
<NONE>; 150MM ROW1G GREY RUNNING OUTLET GUTTER - HALF ROUND RAINWATER	EACH	14.56

BASIC PRICES : LABOUR, PLANT AND MATERIALS

R:PLUMBING : WORKS OF ALTERATION/SMALL WORKS/REPAIR
RW & SOIL PLASTICS

MARLEY: RAINWATER

<NONE>; 150MM RUW1G GREY GUTTER UNION HALF ROUND RAINWATER SYSTEM	EACH	7.50
<NONE>; 68MM RS25G GREY DPPE SHOE DOWNPIPE FITTING	EACH	6.69

BASIC PRICES : LABOUR, PLANT AND MATERIALS

S:MECHANICAL SERVICES - PIPE SUPPLY SYSTEMS : WORKS OF ALTERATION/SMALL WORKS/REPAIR
AIR ELIMINATORS
BROWNALLS AIR ELIMS

<NONE>; BOSS AUTO AIR ELIMINATOR EV C/W 15MM ISO VALVE	EACH	123.12
<NONE>; TYPE A 734S AIR ELIMINATOR	EACH	111.60
<NONE>; TYPE C 736S AIR ELIMINATOR	EACH	135.25

BOSS VALVES
963-965S BALL VALVES

<NONE>; 15MM BOSS 965S BZE BALL VALVE BSP TAPER THRD.F/BORE (1300)	EACH	17.13
<NONE>; 20MM BOSS 965S BZE BALL VALVE BSP TAPER THRD.F/BORE (1300)	EACH	24.21
<NONE>; 25MM BOSS 965S BZE BALL VALVE BSP TAPER THRD.F/BORE (1300)	EACH	27.64
<NONE>; 32MM BOSS 965S BZE BALL VALVE BSP TAPER THRD.F/BORE (1300)	EACH	47.79
<NONE>; 40MM BOSS 965S BZE BALL VALVE BSP TAPER THRD.F/BORE (1300)	EACH	80.23
<NONE>; 50MM BOSS 965S BZE BALL VALVE BSP TAPER THRD.F/BORE (1300)	EACH	147.48

966-968S BALL VALVES

<NONE>; 15MM 966T BOSS NP BRASS B/VVE F/BORE BSPT B.GAS APPR (R751)	EACH	9.42
<NONE>; 20MM 966T BOSS NP BRASS B/VVE F/B BSPT/WRAS. EN331 (R951)	EACH	13.03
<NONE>; 25MM 966T BOSS NP BRASS B/VVE F/B BSPT/WRAS. EN331 (R951)	EACH	18.05
<NONE>; 28MM 968S BOSS NP BRASS B/VVE COMP ENDS B.GAS.APP.R258CX106	EACH	20.45

BOSS BRONZE VENTURI

<NONE>; 15MM 900S BOSS VENTURI VVE. SCREWED BSP	EACH	39.33
<NONE>; 20MM 900S BOSS VENTURI VVE.BSP SCREWED BSP	EACH	65.52
<NONE>; 25MM 900S BOSS VENTURI VVE.BSP SCREWED BSP	EACH	79.50
<NONE>; 32MM 900S BOSS VENTURI VVE.BSP SCREWED BSP	EACH	104.73
<NONE>; 40MM 900S BOSS VENTURI VVE.BSP SCREWED BSP	EACH	139.17
<NONE>; 50MM 900S BOSS VENTURI VVE.BSP SCREWED BSP	EACH	201.18

BOSS DRAIN COCKS - Discontinued

BASIC PRICES : LABOUR, PLANT AND MATERIALS

S:MECHANICAL SERVICES - PIPE SUPPLY SYSTEMS : WORKS OF ALTERATION/SMALL WORKS/REPAIR
BOSS VALVES

BOSS DRAIN COCKS - Discontinued

<NONE>; 15MM(1/2") BOSS 81HU GLANDCOCK 130017 (WRENCH IS 88010818	EACH	29.02
<NONE>; 20MM"BOSS"81HU BZE GLAND COCK COMPLETE WITH LEVER 130008	EACH	47.41
<NONE>; 25MM"BOSS"81HU BZE GLAND COCK COMPLETE WITH LEVER 130009	EACH	67.67

BOSS DUCTILE VENTURI

<NONE>; 100MM 900XSS BOSS VENTURI FLGD PN16 SHORT PATTERN.	EACH	793.56
<NONE>; 100MM 901XS BOSS DRV FLGD PN16 DOUBLE REG VVE	EACH	412.08
<NONE>; 65MM 900XSS BOSS VENTURI FLGD PN16 SHORT PATTERN.	EACH	521.22
<NONE>; 65MM 901XS BOSS DRV FLGD PN16 DOUBLE REG VVE	EACH	302.58
<NONE>; 80MM 900XSS BOSS VENTURI FLGD PN16 SHORT PATTERN.	EACH	618.12
<NONE>; 80MM 901XS BOSS DRV FLGD PN16 DOUBLE REG VVE	EACH	317.43

BOSS GLOBE/CHECK VVS

<NONE>; 15MM 96S BZE HORIZ CHECK VALVE SPRING LOADED PN32 RD PTFE	EACH	51.58
<NONE>; 15MM BOSS 62S BZE.GLOBE VALVE PN32 GLASS FILLED PTFE DISC	EACH	58.37
<NONE>; 20MM 96S BZE HORIZ CHECK VALVE SPRING LOADED PN32 RD PTFE	EACH	65.90
<NONE>; 25MM 96S BZE HORIZ CHECK VALVE SPRING LOADED PN32 RD PTFE	EACH	90.39
<NONE>; 32MM 96S BZE HORIZ CHECK VALVE SPRING LOADED PN32 RD PTFE	EACH	142.57
<NONE>; 40MM 96S BZE HORIZ CHECK VALVE SPRING LOADED PN32 RD PTFE	EACH	168.66
<NONE>; 50MM 99S BZE HORIZ CHECK VALVE PN32 METAL SEAT.	EACH	251.39

BOSS TEST POINTS

<NONE>; 15MM "BOSS" ST15 TEST POINT STANDARD LENGTH	EACH	10.97

VALVES: BOSS MINI

<NONE>; 15MM CUXCU CP/DZR BOSS MINIBALL VVE CU COMPR "WRC"	EACH	11.10
<NONE>; 22MM CUXCU CP/DZR BOSS MINIBALL VVE CU COMPR "WRC"	EACH	19.85

BASIC PRICES : LABOUR, PLANT AND MATERIALS

S:MECHANICAL SERVICES - PIPE SUPPLY SYSTEMS : WORKS OF ALTERATION/SMALL WORKS/REPAIR
BS EN10241 FITTINGS

BS EN10241 FITTINGS

Description	Unit	Price
BLACK MS; 100MM (4") BLK MS CAP 16W	EACH	215.04
BLACK MS; 100MM (4") BLK MS SOCKET 14W	EACH	49.61
BLACK MS; 100MM (4") BLK MS TEE 12W	EACH	713.34
BLACK MS; 15MM (1/2") BLK MS CAP 16W	EACH	8.16
BLACK MS; 15MM (1/2") BLK MS SOCKET 14W	EACH	2.58
BLACK MS; 15MM (1/2") BLK MS TEE 12W	EACH	21.21
BLACK MS; 20MM (3/4") BLK MS SOCKET 14W	EACH	2.97
BLACK MS; 20MM (3/4") BLK MS TEE 12W	EACH	26.08
BLACK MS; 25MM (1") BLK MS CAP 16W	EACH	15.79
BLACK MS; 25MM (1.") BLK MS SOCKET 14W	EACH	3.84
BLACK MS; 25MM (1.") BLK MS TEE 12W	EACH	42.03
BLACK MS; 32MM (1.1/4") BLK MS CAP 16W	EACH	23.45
BLACK MS; 32MM (1.1/4") BLK MS SOCKET 14W	EACH	5.65
BLACK MS; 32MM (1.1/4") BLK MS TEE 12W	EACH	79.45
BLACK MS; 40MM (1.3/4") BLK MS CAP 16W	EACH	26.86
BLACK MS; 40MM (1.3/4") BLK MS SOCKET 14W	EACH	7.23
BLACK MS; 40MM (1.3/4") BLK MS TEE 12W	EACH	79.45
BLACK MS; 50MM (1.") BLK MS TEE 12W	EACH	112.03
BLACK MS; 50MM (2") BLK MS CAP 16W	EACH	39.49
BLACK MS; 50MM (2.") BLK MS SOCKET 14W	EACH	10.85
BLACK MS; 65MM (1.1/2") BLK MS TEE 12W	EACH	409.49
BLACK MS; 65MM (2.1/2") BLK MS CAP 16W	EACH	82.68
BLACK MS; 65MM (2.1/2") BLK MS SOCKET 14W	EACH	18.59
BLACK MS; 80MM (3") BLK MS CAP 16W	EACH	123.87
BLACK MS; 80MM (3") BLK MS SOCKET 14W	EACH	25.99
BLACK MS; 80MM (3") BLK MS TEE 12W	EACH	444.29
GALVANISED; 100MM (4") 14WG GALV MS SOCKET	EACH	66.98
GALVANISED; 100MM (4") GALV MS CAP 17WG	EACH	234.63
GALVANISED; 100MM (4.") GALV MS TEE 12WG	EACH	929.47
GALVANISED; 15MM (1/2") GALV MS CAP 16WG	EACH	10.02
GALVANISED; 15MM (1/2") GALV MS SOCKET 14WG	EACH	3.49
GALVANISED; 15MM (1/2") GALV MS TEE 12WG	EACH	28.63
GALVANISED; 20MM (3/4") GALV MS CAP 16WG	EACH	13.28
GALVANISED; 20MM (3/4") GALV MS SOCKET 14WG	EACH	4.03

© NSR 2008-2009

BASIC PRICES : LABOUR, PLANT AND MATERIALS

S:MECHANICAL SERVICES - PIPE SUPPLY SYSTEMS : WORKS OF ALTERATION/SMALL WORKS/REPAIR
BS EN10241 FITTINGS

BS EN10241 FITTINGS

GALVANISED; 20MM (3/4") GALV MS TEE 12WG	EACH	35.21
GALVANISED; 25MM (1") 14WG GALV MS SOCKET	EACH	5.19
GALVANISED; 25MM (1") GALV MS CAP 16WG	EACH	21.32
GALVANISED; 25MM (1.") GALV MS TEE 12WG	EACH	56.74
GALVANISED; 32MM (1.1/4") GALV MS CAP 16WG	EACH	29.34
GALVANISED; 32MM (1.1/4") GALV MS SOCKET 14WG	EACH	7.63
GALVANISED; 32MM (1.1/4") GALV MS TEE 12WG	EACH	103.38
GALVANISED; 40MM (1.1/2") GALV MS CAP 16WG	EACH	36.26
GALVANISED; 40MM (1.1/2") GALV MS SOCKET 14WG	EACH	9.77
GALVANISED; 40MM (1.1/2") GALV MS TEE 12WG	EACH	103.38
GALVANISED; 50MM (2") 14WG GALV MS SOCKET	EACH	14.65
GALVANISED; 50MM (2") GALV MS CAP 16WG	EACH	50.07
GALVANISED; 50MM (2.") GALV MS TEE 12WG	EACH	151.24
GALVANISED; 65MM (2.1/2") GALV MS CAP 17WG	EACH	83.41
GALVANISED; 65MM (2.1/2") GALV MS SOCKET 14WG	EACH	25.09
GALVANISED; 65MM (2.1/2") GALV MS TEE 12WG	EACH	532.22
GALVANISED; 80MM (3") 14WG GALV MS SOCKET	EACH	35.08
GALVANISED; 80MM (3") GALV MS CAP 16WG	EACH	156.00
GALVANISED; 80MM (3.") GALV MS TEE 12WG	EACH	577.46

BS EN10241 TUBULARS

<NONE>; 100MM(4) BLK MED BARREL NIPPLE 6W	EACH	55.23
<NONE>; 15MM(1/2) BLK HVY LONGSCREW ONLY 300-600MM LONG BSS FIG 4W	EACH	28.64
<NONE>; 20MM(3/4) BLK HVY LONGSCREW LONG 300-600MM LONG BSS FIG 4W	EACH	34.08
<NONE>; 25MM(1) BLK HVY LONGSCREW ONLY 300-600MM LONG BSS FIG 4W	EACH	48.71
<NONE>; 32MM(1.1/4) BLK HVY LONGSCREW ONLY 300-600MM LONG BSS FIG 4W	EACH	61.03
<NONE>; 40MM(1.1/2) BLK HVY BARREL NIP 6W	EACH	7.33
<NONE>; 40MM(1.1/2) BLK HVY LONGSCREW LONG 300-600MM LONG BSS FIG 4W	EACH	72.37
<NONE>; 50MM(2) BLK HVY LONGSCREW ONLY 300-600MM LONG BSS FIG 4W	EACH	103.31
<NONE>; 65MM(2.1/2) BLK HVY LONGSCREW ONLY 300-600MM LONG BSS FIG 4W	EACH	213.34

STEEL FITTINGS BSP

BASIC PRICES : LABOUR, PLANT AND MATERIALS

S:MECHANICAL SERVICES - PIPE SUPPLY SYSTEMS : WORKS OF ALTERATION/SMALL WORKS/REPAIR
BS EN10241 FITTINGS

STEEL FITTINGS BSP

<NONE>; 1" BOSS 40R STEEL UNION BSP FEM/FEM ELECTRO PLATED	EACH	36.75
<NONE>; 1.1/2" BOSS 40R STEEL UNION BSP FEM/FEM ELECTRO PLATED	EACH	75.17
<NONE>; 1.1/4" BOSS 40R STEEL UNION BSP FEM/FEM ELECTRO PLATED	EACH	64.37
<NONE>; 1/2" BOSS 40R STEEL UNION BSP FEM/FEM ELECTRO PLATED	EACH	25.08
<NONE>; 100MM (4") BLK MS HEX NIPPLE	EACH	258.00
<NONE>; 15MM (1/2") BLK MS HEX NIPPLE	EACH	5.44
<NONE>; 15MM (1/2") GALV MS HEX NIPPLE	EACH	7.35
<NONE>; 2" BOSS 40R STEEL UNION BSP FEM/FEM ELECTRO PLATED	EACH	93.47
<NONE>; 25MM (1") GALV MS HEX NIPPLE	EACH	16.87
<NONE>; 3/4" BOSS 40R STEEL UNION BSP FEM/FEM ELECTRO PLATED	EACH	30.56
<NONE>; 32X25MM (1.1/4X1") BLK MS BUSH	EACH	21.52
<NONE>; 40X20MM(1.1/2X3/4)BLK HEX NIPP	EACH	36.55

BS1387 TUBE

BS1387 TUBE P/E

<NONE>; 100MM (4") RED HVY P/E TUBE FROM CORUS	PER METRE	33.85
<NONE>; 15MM (1/2") RED HVY P/E TUBE FROM CORUS	PER METRE	4.19
<NONE>; 20MM (3/4") RED HVY P/E TUBE FROM CORUS	PER METRE	4.77
<NONE>; 25MM (1") RED HVY P/E TUBE FROM CORUS	PER METRE	6.97
<NONE>; 32MM (1.1/4") RED HVY P/E TUBE FROM CORUS	PER METRE	8.65
<NONE>; 40MM (1.1/2") RED HVY P/E TUBE FROM CORUS	PER METRE	10.08
<NONE>; 50MM (2") RED HVY P/E TUBE FROM CORUS	PER METRE	14.01
<NONE>; 65MM (2.1/2") RED HVY P/E TUBE FROM CORUS	PER METRE	19.05
<NONE>; 80MM (3") RED HVY P/E TUBE FROM CORUS	PER METRE	24.25

BS1387 TUBE P/E GLVD

<NONE>; 100MM(4) GALV MED P/E TUBE BS1387 FROM CORUS	PER METRE	38.16
<NONE>; 15MM (1/2") GALV MED P/E TUBE FROM CORUS	PER METRE	5.22
<NONE>; 20MM (3/4") GALV MED P/E TUBE FROM CORUS	PER METRE	5.45
<NONE>; 25MM (1") GALV MED P/E TUBE FROM CORUS	PER METRE	7.61
<NONE>; 32MM (1.1/4") GALV MED P/E TUBE FROM CORUS	PER METRE	9.42

BASIC PRICES : LABOUR, PLANT AND MATERIALS

S:MECHANICAL SERVICES - PIPE SUPPLY SYSTEMS : WORKS OF ALTERATION/SMALL WORKS/REPAIR
BS1387 TUBE
BS1387 TUBE P/E GLVD

<NONE>; 40MM (1.1/2") GALV MED P/E TUBE FROM CORUS	PER METRE	10.95
<NONE>; 50MM (2") GALV MED P/E TUBE FROM CORUS	PER METRE	15.36
<NONE>; 65MM(2.1/2) GALV MED P/E TUBE FROM CORUS	PER METRE	20.84
<NONE>; 80MM (3") GALV MED P/E TUBE BS1387 FROM CORUS	PER METRE	26.99

BS1965 FITTINGS
BS1965 WELD FITTINGS

<NONE>; 100MM (4") MED ELBOW LR 90 BS1965	EACH	42.64
<NONE>; 100MM (4") MS CAP	EACH	80.04
<NONE>; 100MM (4") MS TEE	EACH	170.16
<NONE>; 15MM (1/2") MED ELBOW LR 90 BS 1965	EACH	9.40
<NONE>; 15MM (1/2") MS CAP	EACH	43.56
<NONE>; 15MM (1/2") MS TEE	EACH	88.90
<NONE>; 20MM (3/4") MED ELBOW LR 90 BSD 1965	EACH	9.20
<NONE>; 20MM (3/4") MS CAP	EACH	43.56
<NONE>; 20MM (3/4") MS TEE	EACH	88.90
<NONE>; 25MM (1") MED ELBOW LR 90 BS 1965	EACH	12.22
<NONE>; 25MM (1") MS CAP	EACH	43.56
<NONE>; 25MM (1") MS TEE	EACH	88.90
<NONE>; 32MM (1.1/4") MED ELBOW LR 90 BS 1965	EACH	14.58
<NONE>; 32MM (1.1/4") MS CAP	EACH	43.56
<NONE>; 32MM (1.1/4") MS TEE	EACH	88.90
<NONE>; 40MM (1.1/2") MED ELBOW LR 90 BS1965	EACH	14.74
<NONE>; 40MM (1.3/4") MS CAP	EACH	43.82
<NONE>; 40MM (1.3/4") MS TEE	EACH	88.90
<NONE>; 50MM (2") MED ELBOW LR 90 BS1965	EACH	20.16
<NONE>; 50MM (2") MS TEE	EACH	93.60
<NONE>; 50MM (2.") MS CAP	EACH	51.20
<NONE>; 65MM (1.1/4") MS CAP	EACH	60.14
<NONE>; 65MM (2.1/2") MED ELBOW LR 90 BS1965	EACH	25.96
<NONE>; 65MM (2.1/2") MS TEE	EACH	136.54
<NONE>; 80MM (3") MED ELBOW LR 90 BS1965	EACH	28.18
<NONE>; 80MM (3") MS CAP	EACH	61.20

BASIC PRICES : LABOUR, PLANT AND MATERIALS

S:MECHANICAL SERVICES - PIPE SUPPLY SYSTEMS : WORKS OF ALTERATION/SMALL WORKS/REPAIR

BS1965 FITTINGS

BS1965 WELD FITTINGS

<NONE>; 80MM (3") MS TEE	EACH	136.76

CONSUMABLES

MISC COMPOUNDS

<NONE>; RADIATOR AIR BLEEDER (COMPACT PLUG & AIR VENT)	EACH	8.30

COPPER FTGS+VVE

COMP: CONEX A

<NONE>; 15MM FIG 301 STRAIGHT COUPLER	EACH	1.82
<NONE>; 15X15 S301 ST CPLER DZR	EACH	3.28
<NONE>; 22MM FIG 301 STRAIGHT COUPLER	EACH	3.15
<NONE>; 22X .75 S302 ST CPLER / IRON DZR	EACH	4.67
<NONE>; 28MM X 1 FIG 302 ST COUPLER	EACH	5.38

COMP: CONEX MAIN

<NONE>; 108X108MM S301 ST COUPLER DZR	EACH	585.00
<NONE>; 35MM FIG 301 STRAIGHT COUPLER	EACH	25.76
<NONE>; 35MM FIG S601EQ EQUAL TEE	EACH	44.87
<NONE>; 35MMX 1 S303 STRT COUPLER	EACH	27.75
<NONE>; 54MM 301/S301 STRAIGHT COUPLER	EACH	51.22
<NONE>; 67X67MM S301 ST CPLER DZR	EACH	238.54
<NONE>; 76X76MM S301 ST CPLER DZR	EACH	359.54

COMP: KUT MAIN

<NONE>; 15MM 610AC STRT COUPLING AIR RELEASE VALVE	EACH	12.90
<NONE>; 22MM 610AC STRT COUPLING WITH AIR RELEASE VALVE	EACH	17.03

COMP: KUTERLITE KS

<NONE>; 15X.5 912 FEMALE COUPLING	EACH	3.31

DELCOP MAIN

<NONE>; 28MM 601 STRT COUPLING	EACH	1.77
<NONE>; 35MM 601 STRT COUPLING	EACH	6.34
<NONE>; 42MM 601 STRT COUPLING	EACH	10.28
<NONE>; 54MM 601 STRT COUPLING	EACH	18.75

BASIC PRICES : LABOUR, PLANT AND MATERIALS

S:MECHANICAL SERVICES - PIPE SUPPLY SYSTEMS : WORKS OF ALTERATION/SMALL WORKS/REPAIR
COPPER FTGS+VVE

MISC BRASS

<NONE>; 15MM 885S BRASS F/F STOPCOCK BS1010	EACH	16.45
<NONE>; 20MM 885S BRASS F/F STOPCOCK BS1010	EACH	17.74
<NONE>; 25MM 885S BRASS F/F STOPCOCK BS1010	EACH	34.58
<NONE>; 32MM 885S BRASS F/F STOPCOCK BS1010	EACH	63.60
<NONE>; 40MM 885S BRASS F/F STOPCOCK BS1010	EACH	74.19
<NONE>; 50MM 885S BRASS F/F STOPCOCK BS1010	EACH	92.59

SOLDER RING:YORK GEN

<NONE>; 22MM YP69F UNION CONN	EACH	18.19
<NONE>; 28MM YP1 COUPLING	EACH	1.92
<NONE>; 28MM YP24 TEE	EACH	8.55
<NONE>; 42MM YP1 COUPLING	EACH	10.44
<NONE>; 54MM YP1 COUPLING	EACH	19.25

SOLDER RING:YORK YPS

<NONE>; 15MM YPS12 ELBOW	EACH	0.58
<NONE>; 15MM YPS24 TEE	EACH	1.09
<NONE>; 22MM YPS1 COUPLING	EACH	0.85
<NONE>; 22MM YPS24 TEE	EACH	3.47

VALVES: BIBCOCKS

<NONE>; 1/2"BS1010 HOSE UNION BIBCOCK 59.07.733	EACH	5.76
<NONE>; 3/4"BS1010 HOSE UNION BIBCOCK 59.07.832	EACH	11.46

VALVES: DRAIN OFF

<NONE>; 1/2" BRASS RAD AIRVENT PLUG (BBL860)	EACH	1.65
<NONE>; 15MM SOLDER IN DRAIN COCK TYPE A	EACH	1.08
<NONE>; 15MM TYPE A SOLDER DRAIN COCK PLEASE USE 85510519	EACH	1.83

VALVES: ISO/SERVICE

<NONE>; 15MM 3100/BC BRASS ISOLATING VALVE 41.20.136	EACH	1.99
<NONE>; 22MM 3100/BD BRASS ISOLATING VALVE 41.20.181	EACH	3.87

COPPER TUBE

TUBE: LAW:35-159 T X

<NONE>; 54X1.2X3M CU-TUBE LAWTON (TX) EN1057 R250 (34)	PER METRE	28.87

BASIC PRICES : LABOUR, PLANT AND MATERIALS

S:MECHANICAL SERVICES - PIPE SUPPLY SYSTEMS : WORKS OF ALTERATION/SMALL WORKS/REPAIR
COPPER TUBE

TUBE: WTF: 15-28 T X

<NONE>; 15X0.7X6M CU-TUBE (TX) EN1057 R250 FORMERLY BS2871 TX	PER METRE	6.72
<NONE>; 22X0.9X2M CU-TUBE (TX) EN1057 R250 FORMERLY BS2871 TX	PER METRE	13.44
<NONE>; 22X0.9X3M CU-TUBE (TX) EN1057 R250 FORMERLY BS2871 TX	PER METRE	13.44
<NONE>; 22X0.9X6M CU-TUBE (TX) EN1057 R250 FORMERLY BS2871 TX	PER METRE	13.44
<NONE>; 28X0.9X6M CU-TUBE (TX) EN1057 R250 FORMERLY BS2871 TX	PER METRE	16.95

TUBE: WTF: 6-10 T W

<NONE>; 10X0.7X10M CU-TUBE (TW) EN1057 R220 FORMERLY BS2871 TW	PER METRE	6.25

TUBE: WTF:35-159 T X

<NONE>; 108X1.5X6M CU-TUBE (TX) EN1057 R250 FORMERLY BS2871 TX	PER METRE	162.67
<NONE>; 35X1.2X6M CU-TUBE (TX) EN1057 R250 FORMERLY BS2871 TX	PER METRE	39.37
<NONE>; 42X1.2X6M CU-TUBE (TX) EN1057 R250 FORMERLY BS2871 TX	PER METRE	47.88
<NONE>; 54X1.2X6M CU-TUBE (TX) EN1057 R250 FORMERLY BS2871 TX	PER METRE	61.60
<NONE>; 67X1.2X6M CU-TUBE (TX) EN1057 R250 FORMERLY BS2871 TX	PER METRE	79.88
<NONE>; 76X1.5X6M CU-TUBE (TX) EN1057 R250 FORMERLY BS2871 TX	PER METRE	113.11

TUBE: YCT: 15-28 T.X

<NONE>; 15X0.7X2M CU-TUBE (TX) EN1057 R250 FORMERLY BS2871 TX	PER METRE	6.72
<NONE>; 15X0.7X6M CU-TUBE (TX) EN1057 R250 FORMERLY BS2871 TX	PER METRE	6.72
<NONE>; 22X0.9X3M CU-TUBE (TX) EN1057 R250 FORMERLY BS2871 TX	PER METRE	13.44
<NONE>; 22X0.9X6M CU-TUBE (TX) EN1057 R250 FORMERLY BS2871 TX	PER METRE	13.44
<NONE>; 28X0.9X6M CU-TUBE (TX) EN1057 R250 FORMERLY BS2871 TX	PER METRE	16.95

TUBE: YCT: REMAINDER

<NONE>; 15X0.7X6M CP CU-TUBE (TX) EN1057 R250	PER METRE	12.01

TUBE: YCT:35-159 T.X

BASIC PRICES : LABOUR, PLANT AND MATERIALS

S:MECHANICAL SERVICES - PIPE SUPPLY SYSTEMS : WORKS OF ALTERATION/SMALL WORKS/REPAIR
COPPER TUBE

TUBE: YCT:35-159 T.X

<NONE>; 35X1.2X6M CU-TUBE (TX) EN1057 R250 FORMERLY BS2871 TX	PER METRE	39.37
<NONE>; 54X1.2X3M CU-TUBE (TX) EN1057 R250 FORMERLY BS2871 TX	PER METRE	61.60

CRANE VALVES

CRANE BRONZE VALVES

<NONE>; 100MM (4") D151 BZE GATE VVE	EACH	731.16
<NONE>; 100MM (4) D156 BRASS GATE VVE	EACH	215.84
<NONE>; 15MM (1/2") D151A DZR GATE VVE WH	EACH	14.36
<NONE>; 15MM (1/2") D4 BZE GLOBE VVE	EACH	22.65
<NONE>; 15MM (1/2") D921 DBLE REG VVE BZE BSP SCREWED	EACH	56.34
<NONE>; 15MM (1/2") D931 FO D/REG VVE BZE BSP SCREWED	EACH	77.28
<NONE>; 15MM(1/2) D297PN25 BZE STRNR SCREEN SUIT OIL&WATER WAS PN32	EACH	35.66
<NONE>; 20MM (3/4") D104 BZE CHECK VVE	EACH	32.11
<NONE>; 20MM (3/4") D151 BZE GATE VVE	EACH	36.51
<NONE>; 20MM (3/4") D151A DZR GATE VVE WH	EACH	16.96
<NONE>; 20MM (3/4") D4 BZE GLOBE VVE	EACH	28.45
<NONE>; 20MM (3/4") D921 DBLE REG VVE BZE BSP SCREWED	EACH	96.55
<NONE>; 20MM (3/4") D931 FO D/REG VVE BZE BSP SCREWED	EACH	124.18
<NONE>; 20MM(3/4) D297PN25 BZE STRNR SCREEN SUIT OIL&WATER-WASPN32	EACH	45.30
<NONE>; 25MM (1") D104 BZE CHECK VVE	EACH	47.16
<NONE>; 25MM (1") D151 BZE GATE VVE	EACH	47.25
<NONE>; 25MM (1") D151A DZR GATE VVE WH	EACH	23.42
<NONE>; 25MM (1") D4 BZE GLOBE VVE	EACH	43.15
<NONE>; 25MM (1") D921 DBLE REG VVE BZE BSP SCREWED	EACH	121.13
<NONE>; 25MM (1") D931 FO D/REG VVE BZE BSP SCREWED	EACH	147.68
<NONE>; 25MM(1) D297PN25 BZE STRNR SCREEN SUIT OIL&WATER-WASPN32	EACH	60.09
<NONE>; 32MM (1.1/4") D104 BZE CHECK VVE	EACH	71.82
<NONE>; 32MM (1.1/4") D151 BZE GATE VVE	EACH	67.41
<NONE>; 32MM (1.1/4") D151A DZR GATE VVW WH	EACH	35.09

BASIC PRICES : LABOUR, PLANT AND MATERIALS

S:MECHANICAL SERVICES - PIPE SUPPLY SYSTEMS : WORKS OF ALTERATION/SMALL WORKS/REPAIR
CRANE VALVES

CRANE BRONZE VALVES

<NONE>; 32MM (1.1/4") D4 BZE GLOBE VV$	EACH	65.49
<NONE>; 32MM (1.1/4) D921 DBLE REG VVE BZE BSP SCREWED	EACH	169.60
<NONE>; 32MM (1.1/4) D931 FO D/REG VVE BZE BSP SCREWED	EACH	198.01
<NONE>; 32MM(1.1/4) D297PN25 BZE STRNR SCREEN SUIT OIL&WATER-WASPN32	EACH	97.05
<NONE>; 40MM (1.1/2") D104 BZE CHECK VVE	EACH	93.58
<NONE>; 40MM (1.1/2") D151 BZE GATE VVE	EACH	92.27
<NONE>; 40MM (1.1/2") D151A DZR GATE VVE WH	EACH	49.42
<NONE>; 40MM (1.1/2") D4 BZE GLOBE VVE	EACH	85.63
<NONE>; 40MM (1.1/2) D921 DBLE REG VVE BZE BSP SCREWED	EACH	244.05
<NONE>; 40MM (1.1/2) D931 FO D/REG VVE BZE BSP SCREWED	EACH	287.15
<NONE>; 40MM(1.1/2) D297PN25 BZE STRNR SCREEN SUIT OIL&WATER-WAS PN3	EACH	125.53
<NONE>; 50MM (2") D104 BZE CHECK VVE	EACH	140.23
<NONE>; 50MM (2") D151 BZE GATE VVE	EACH	131.85
<NONE>; 50MM (2") D151A DZR GATE VVE WH	EACH	68.46
<NONE>; 50MM (2") D4 BZE GLOBE VVE	EACH	130.26
<NONE>; 50MM (2") D921 DBLE REG VVE BZE BSP SCREWED	EACH	357.09
<NONE>; 50MM (2") D931 FO D/REG VVE BZE BSP SCREWED	EACH	410.75
<NONE>; 50MM(2) D297PN25 BZE STRAINER SCREEN SUIT OIL&WATER-WASPN32	EACH	210.26
<NONE>; 65MM (2.1/2") D151A DZR GATE VVE WH	EACH	120.09
<NONE>; 65MM (2.1/2") D7 BZE GLOBE VVE	EACH	690.26
<NONE>; 80MM (3") D151A DZR GATE VVE WH	EACH	174.86
<NONE>; 80MM (3") D7 BZE GLOBE VVE	EACH	967.86

CRANE METRIC CI VVES

<NONE>; 100MM (4") DM921 CI DR VVE FGD PN16	EACH	859.74
<NONE>; 100MM (4") DM931 CI FM DR VVE	EACH	971.31
<NONE>; 100MM (4") FM369 PN16 CI GLOVE VVE	EACH	734.34
<NONE>; 100MM FM276 CI STRAINER SS SCRN 0.75MM WATER GAS STEAM	EACH	387.72
<NONE>; 65MM (2.1/2") DM921 CI DR VVE FGD PN16	EACH	511.83
<NONE>; 65MM (2.1/2") DM931 CI FM DR VVE	EACH	609.72

BASIC PRICES : LABOUR, PLANT AND MATERIALS

S:MECHANICAL SERVICES - PIPE SUPPLY SYSTEMS : WORKS OF ALTERATION/SMALL WORKS/REPAIR
CRANE VALVES
CRANE METRIC CI VVES

<NONE>; 65MM FM276 CI STRAINER SS SCRN 0.75MM WATER GAS STEAM	EACH	223.87
<NONE>; 80MM (3") DM921 CI DR VVE FGD PN16	EACH	626.48
<NONE>; 80MM (3") DM931 CI FM DR VVE	EACH	737.35
<NONE>; 80MM FM276 CI STRAINER SS SCRN 0.75MM WATER GAS STEAM	EACH	265.00

CRANE QTR TURN VVES

<NONE>; 100MM (4") D191 BRASS BALL VVE BSPT	EACH	473.92
<NONE>; 15MM (1/2") D191 BRASS BALL VVE BSPT	EACH	9.95
<NONE>; 15MM (1/2") D340 BZE COCK	EACH	20.36
<NONE>; 15MM (1/2") D344.5 BZE DRAW OFF COCK	EACH	44.96
<NONE>; 20MM (3/4") D191 BRASS BALL VVE BSPT	EACH	12.49
<NONE>; 20MM (3/4") D340 BZE D/O COCK	EACH	31.22
<NONE>; 20MM (3/4") D344.5 BZE DRAW $ OFF COCK	EACH	60.91
<NONE>; 25MM (1") D191 BRASS BALL VVE BSPT	EACH	18.28
<NONE>; 25MM (1") D340 BZE D/O COCK	EACH	91.53
<NONE>; 25MM (1") D344.5 BZE DRAW OFF COCK	EACH	91.31
<NONE>; 32MM (1.1/4") D191 BRASS BALL VVE BSPT	EACH	28.03
<NONE>; 40MM (1.1/2") D191 BRASS BALL VVE BSPT	EACH	40.67
<NONE>; 50MM (2") D191 BRASS BALL VVE BSPT	EACH	63.31
<NONE>; 65MM (2.1/2") D191 BRASS BALL VVE BSPT	EACH	144.20
<NONE>; 80MM (3") D191 BRASS BALL VVE BSPT	EACH	226.65

GAUGES
GAUGE SYPHONS

<NONE>; 8MM 394S BRASS GAUGE COCK MAX.PRESSURE 150PSI (0201BH)	EACH	31.87
<NONE>; 8MM 400S BRASS RING SYPHON	EACH	91.26
<NONE>; 8MM 401S MS U SYPHON	EACH	11.27
<NONE>; 8MM 403S MS RING SYPHON	EACH	16.43

GAUGES & THERMOMETER

<NONE>; 1/4"BSP BRASS PRESS GAUGE VVE M/F RATING 125BAR AT 120DEG C	EACH	43.32
<NONE>; 100MM 408DM P/GAUGE 7BAR/100LB STEEL CASE	EACH	37.26
<NONE>; 100MM 415DM ALT GAUGE 60M/200 BRASS CASE	EACH	89.13

BASIC PRICES : LABOUR, PLANT AND MATERIALS

S:MECHANICAL SERVICES - PIPE SUPPLY SYSTEMS : WORKS OF ALTERATION/SMALL WORKS/REPAIR

GAUGES

GAUGES & THERMOMETER

Description	Unit	Rate
<NONE>; 100MM 435S MERC/ST THERM 0/120 DEG C&F BLK CASE CHORME BEZEL	EACH	165.50
<NONE>; 100MM 438S VP THERM.20/120C&F BLK.CASE ST.STEM BOTTOM ENTRY	EACH	112.18
<NONE>; 100MM 439S VP THERM.20/120C&F BLK.CASE ANGLE STEM B/ENTRY	EACH	112.20
<NONE>; 150MM 408DM P/GAUGE 2BAR/30LB STEEL CASE	EACH	65.84
<NONE>; 150MM 415DM ALT GAUGE 40M/130F BRASS CASE	EACH	85.59
<NONE>; 150MM 435S MERC/ST THERM 0/120 DEG C&F BLK CASE CHROME BEZEL	EACH	186.25
<NONE>; 200MM 271B ST SIKA THERM 0-120	EACH	91.84
<NONE>; 200MM 272B ANGLE SIKA THERM	EACH	94.76

GENERAL

Sundries

Description	Unit	Rate
Materials; Unit material rate	BASE	1.35

HATTERSLEY VVES

HATTS BRONZE VALVES

Description	Unit	Rate
<NONE>; 100MM(4") 35E BZE GATE VALVE BSTE D/D IMP FLGD - HATTERSLEY	EACH	1,427.06
<NONE>; 15MM(1/2") 2086TRV BZE D/FLO POL CHROME ANGLE PATT HATTS	EACH	42.23
<NONE>; 15MM(1/2") 42 BZE CHECK VALVE BSP THREADED - HATTERSLEY	EACH	28.52
<NONE>; 15MM(1/2") 47 BZE CHECK VALVE BSP THREADED - HATTERSLEY	EACH	22.13
<NONE>; 15MM(1/2") 5 BZE GLOBE VALVE BSP THREADED - HATTERSLEY	EACH	25.29
<NONE>; 15MM(1/2") 81HU BZE GLAND COCK HOSE UNION - HATTERSLEY	EACH	35.66
<NONE>; 20MM(3/4") 42 BZE CHECK VALVE BSP THREADED - HATTERSLEY	EACH	42.49
<NONE>; 20MM(3/4") 47 BZE CHECK VALVE BSP THREADED - HATTERSLEY	EACH	26.32
<NONE>; 20MM(3/4") 5 BZE GLOBE VALVE BSP THREADED - HATTERSLEY	EACH	34.46
<NONE>; 20MM(3/4") 81HU BZE GLAND COCK HOSE UNION - HATTERSLEY	EACH	51.99
<NONE>; 25MM(1") 42 BZE CHECK VALVE BSP THREADED - HATTERSLEY	EACH	60.42

BASIC PRICES : LABOUR, PLANT AND MATERIALS

S:MECHANICAL SERVICES - PIPE SUPPLY SYSTEMS : WORKS OF ALTERATION/SMALL WORKS/REPAIR

HATTERSLEY VVES

HATTS BRONZE VALVES

<NONE>; 25MM(1") 47 BZE CHECK VALVE BSP THREADED - HATTERSLEY	EACH	36.46
<NONE>; 25MM(1") 5 BZE GLOBE VALVE BSP THREADED - HATTERSLEY	EACH	52.64
<NONE>; 32MM(1.1/4") 42 BZE CHECK VVE BSP THREADED - HATTERSLEY	EACH	81.28
<NONE>; 32MM(1.1/4") 47 BZE CHECK VVE BSP THREADED - HATTERSLEY	EACH	61.78
<NONE>; 32MM(1.1/4") 5 BZE GLOBE VALVE BSP THREADED - HATTERSLEY	EACH	74.29
<NONE>; 40MM(1.1/2") 42 BZE CHECK VVE BSP THREADED - HATTERSLEY	EACH	95.03
<NONE>; 40MM(1.1/2") 47 BZE CHECK VVE BSP THREADED - HATTERSLEY	EACH	76.86
<NONE>; 40MM(1.1/2") 5 BZE GLOBE VALVE BSP THREADED - HATTERSLEY	EACH	92.25
<NONE>; 50MM(2") 42 BZE CHECK VALVE BSP THREADED - HATTERSLEY	EACH	153.34
<NONE>; 50MM(2") 47 BZE CHECK VALVE BSP THREADED - HATTERSLEY	EACH	117.88
<NONE>; 50MM(2") 5 BZE GLOBE VALVE BSP THREADED - HATTERSLEY	EACH	146.14
<NONE>; 65MM(2.1/2") 13 BZE R/D VALVE BSP THREADED - HATTERSLEY	EACH	371.41
<NONE>; 65MM(2.1/2") 47 BZE CHECK VVE BSP THREADED - HATTERSLEY	EACH	240.62
<NONE>; 80MM(3") 13 BZE R/D VALVE BSP THREADED - HATTERSLEY	EACH	538.20
<NONE>; 80MM(3") 47 BZE CHECK VALVE BSP THREADED - HATTERSLEY	EACH	340.23

HATTS CI VALVES

<NONE>; 100MM(4") 549E CI GATE VVE BSTE IMP FLGD - HATTERSLEY	EACH	476.29
<NONE>; 100MM(4") 651E CI CHK VVE BSTE IMP SWING PATT HATTERSLEY	EACH	616.47
<NONE>; 80MM(3") 651E CI CHK VVE BSTE IMP SWING PATT HATTERSLEY	EACH	467.06

HATTS METRIC CI VVES

<NONE>; 100MM(4") M651PN16 CI CHECK VVE METRIC - HATTERSLEY	EACH	366.65

HEAT EXCHANGERS

CYLINDERS/IMMERSION HEATERS

BASIC PRICES : LABOUR, PLANT AND MATERIALS

S:MECHANICAL SERVICES - PIPE SUPPLY SYSTEMS : WORKS OF ALTERATION/SMALL WORKS/REPAIR
HEAT EXCHANGERS
CYLINDERS/IMMERSION HEATERS

	<NONE>; GLASS FIBRE INSUL JACKET - 42 in x 18 in x 80 mm	nr	14.82
	<NONE>; IMMERSION HEATER/STAT - 11IN.	nr	31.65
	<NONE>; IMMERSION HEATER/STAT - 27 IN.	nr	37.95
	<NONE>; TELFORD COPPER CYLINDER IND COMBI LAGGED - 1050 X 450 mm	nr	260.29
	<NONE>; TELFORD COPPER CYLINDER IND COMBI LAGGED - 1200 X 450 mm	nr	294.32
	<NONE>; TELFORD COPPER CYLINDER IND COMBI LAGGED - 900 X 450 mm	nr	248.11
	<NONE>; TELFORD COPPER CYLINDER IND LAGGED - 1050 X 400 mm	nr	155.21
	<NONE>; TELFORD COPPER CYLINDER IND LAGGED - 1200 X 400 mm	nr	213.23
	<NONE>; TELFORD COPPER CYLINDER IND LAGGED - 900 X 400 mm	nr	160.43

HEATING CONTROL
ALCON INCENTIVE ITEM

	<NONE>; GB4C 1/2 GAS VVE FAST OPEN240V	EACH	73.04
	<NONE>; GB6B 3/4 GAS VVE FAST OPEN230V	EACH	87.94
	<NONE>; GB7B 1 GAS VVE FAST OPEN230V	EACH	148.80
	<NONE>; HWA10 2 GAS VVE 240 SLOW OPEN	EACH	879.06
	<NONE>; HWA11 2.1/2" GAS VALVE 230V	EACH	1,151.02
	<NONE>; HWA9 1.1/2 GAS VVE 230V $ SLOW OPEN	EACH	898.95

BOSSMIX VALVES

	<NONE>; 15/22MM BOSSMIX THERM MIXING VALVE FAILSAFE CP DZR BRASS - DISCONTINUED	EACH	70.12
	<NONE>; 15MM BOSS MIX CHECK VALVE KIT CONSISTING OF 2 CP DZR VALVES	EACH	23.30
	<NONE>; 15MM BOSSMIX PLUS SERVICE KIT 46-1012 FOR 86310006 - DISCONTINUED	EACH	50.13
	<NONE>; 15MM BOSSMIXPLUS C/W STRAINER CHECK VVS (S/KIT 86321036)	EACH	127.85
	<NONE>; 22MM BOSS MIX CHECK VALVE KIT CONSISTING OF 2 CP DZR VALVES	EACH	23.30
	<NONE>; 22MM BOSSMIX PLUS SERVICE KIT 46-1013 FOR 86310017 - DISCONTINUED	EACH	52.38
	<NONE>; 22MM BOSSMIXPLUS C/W STRAINER & CHECK VVS (S/KIT 86321047)	EACH	134.75

COMM SHOWER PRODUCTS

BASIC PRICES : LABOUR, PLANT AND MATERIALS

S:MECHANICAL SERVICES - PIPE SUPPLY SYSTEMS : WORKS OF ALTERATION/SMALL WORKS/REPAIR

HEATING CONTROL

COMM SHOWER PRODUCTS

<NONE>; 22MM ISOLATING STRAINER CHECK VALVE (SPVE0016J)	EACH	37.00
<NONE>; 60028CP 28MM INTAMIX PRO THERMOSTATIC MIXING VALVE	EACH	354.22
<NONE>; BATH/SHOWER RIGID RISER	EACH	64.00
<NONE>; CONTROL KNOB ASSY SPKB0009P MEYNELL SM1	EACH	28.00
<NONE>; ER-S 310 SHOWER KIT	EACH	71.00
<NONE>; MIRA 15 CARTRIDGE (902.55)	EACH	65.33
<NONE>; MULTI MODE SHOWER HEAD CP	EACH	34.84
<NONE>; V6 EXPOSED THERMOSTATIC SHOWER MIXER	EACH	199.00
<NONE>; V8/3B RECESSED THERMOSTATIC SHOWER MIXER	EACH	239.00
<NONE>; VR LARGE SHOWER HEAD RISING/CONCEALED SUPPLY	EACH	78.23

DRAYTON TRV

<NONE>; 0705150 15MM ANGLE TRV4	EACH	23.50
<NONE>; 0705154 .75" ANGLE TRV4	EACH	37.27
<NONE>; 0715623 BODY ONLY 1" ANGLE SP TRV4	EACH	49.43
<NONE>; 0725006 TRV4 INTEG HEAD ONLY (WAS 0794006)	EACH	20.25

HNYWELL RESIDENTIAL

<NONE>; LINK TUBE ISOLATING VALVE 10116008	EACH	17.00

L K FIRE EQUIPMENT

<NONE>; 10905502 BREECH BOX VERT FORMERLEY LK 15137)	EACH	371.66
<NONE>; 10905601 LANDING BOX	EACH	371.66
<NONE>; LK15112 2.5 GATE VVE	EACH	315.57
<NONE>; LK15114 65MM NP16 GATE VALVE	EACH	315.57
<NONE>; LK15133 CI INLET X2 4" BSTD	EACH	389.20
<NONE>; LK16116 19MM SW/AUTO REEL	EACH	490.53
<NONE>; LK16128 25MM SW/AUTO REEL	EACH	507.58
<NONE>; LK16214 30M HOSE (19MM)	EACH	168.55
<NONE>; LK16215 35M HOSE (19MM)	EACH	196.55

HOT AND COLD WATER SERVICES

Ball valve and float

BASIC PRICES : LABOUR, PLANT AND MATERIALS

S:MECHANICAL SERVICES - PIPE SUPPLY SYSTEMS : WORKS OF ALTERATION/SMALL WORKS/REPAIR
HOT AND COLD WATER SERVICES

Ball valve and float

Ball valve and float; 15mm, equilibrium pattern, high pressure, plastic float	nr	19.35
Ball valve and float; 15mm, Portsmouth pattern, high pressure, plastic float	nr	6.11
Ball valve and float; 22mm, equilibrium pattern, high pressure, plastic float	nr	29.73
Ball valve and float; 22mm, Portsmouth pattern, high pressure, plastic float	nr	12.89
Ball valve and float; 28mm, equilibrium pattern, high pressure	nr	59.40
Ball valve and float; 50mm, equilibrium pattern, high pressure, flanged connections bronze	nr	306.54
Ball valve and float; 75mm, equilibrium pattern, high pressure, flanged connections bronze	nr	1,188.74
Ball valve and float; 75mm, equilibrium pattern, high pressure, flanged connections cast iron	nr	357.57

Insulation

Insulation; 15mm CWS insulation, exposed	m	7.11

IND/PROC VALVES
KITZ CS/SS FLG BALLV

<NONE>; 100MM REDUCED BORE F/S.S/S BALL VALVE	EACH	662.87
<NONE>; 15MM REDUCED BORE F/S.S/S BALL VALVE	EACH	90.19
<NONE>; 20MM REDUCED BORE F/S.S/S BALL VALVE	EACH	107.80
<NONE>; 25MM REDUCED BORE F/S.S/S BALL VALVE	EACH	131.49
<NONE>; 40MM REDUCED BORE F/S.S/S BALL VALVE	EACH	221.20
<NONE>; 50MM REDUCED BORE F/S.S/S BALL VALVE	EACH	299.94
<NONE>; 80MM REDUCED BORE F/S.S/S BALL VALVE	EACH	467.64

INDUSTR. PUMPS
BOSS S.S. SUMP

<NONE>; BOSS 115 SUMP 3PH	EACH	501.00

BOSS SS SUMP PUMP

<NONE>; BOSS 200 1PH C/W FLOAT 10M CABLE 1 1/4" CONNS	EACH	875.00
<NONE>; BOSS 200 3PH C/W FLOAT 10M CABLE 1 1/4" CONNS	EACH	888.00
<NONE>; BOSS 415 1PH C/W FLOAT 20M CABLE 1 1/4" CONNS	EACH	866.42

BASIC PRICES : LABOUR, PLANT AND MATERIALS

S:MECHANICAL SERVICES - PIPE SUPPLY SYSTEMS : WORKS OF ALTERATION/SMALL WORKS/REPAIR
INDUSTR. PUMPS

BOSS SS SUMP PUMP

<NONE>; BOSS 415 3PH C/W FLOAT 20M CABLE 1 1/4" CONNS	EACH	839.96
<NONE>; BOSS 418 SUMP 1PH NO FLOAT 1 1/4" CONN 2.5HP	EACH	787.05
<NONE>; BOSS 422 1PH C/W FLOAT 20M CABLE 1 1/4" CONNS	EACH	1,111.13
<NONE>; BOSS 422 3PH C/W FLOAT 20M CABLE 1 1/4" CONNS	EACH	1,084.67
<NONE>; WMTM 15A 220/240VAC 0.22KW C/W FLOAT SWITCH 5M CABLE	EACH	234.13
<NONE>; WMTM 17A 220/240VAC 0.45KW C/W FLOAT SWITCH 5M CABLE	EACH	304.24
<NONE>; WMTM 55A 220/240VAC 0.6KW C/W FLOAT SWITCH 5M CABLE	EACH	383.60
<NONE>; WMTM 58A 220/240VAC 0.75KW C/W FLOAT SWITCH 5M CABLE	EACH	403.45

BROWNALL LABTAPS

<NONE>; 152MM REACH XL1219C BLACK BAL FLOW SWANNECK MIXER TAP	EACH	89.39
<NONE>; 152MM REACH XL1401C-8M84 BLACK BAL FLOW WRIST ACTION SWANNEC	EACH	139.61
<NONE>; BROWNALL BLACK XL1204CY	EACH	48.68
<NONE>; XL1207D-8M41 BLACK SWANNECK $ SIDE ARM BIBTAP B.W.F	EACH	102.12

FLYGT SUBMERS PUMPS

<NONE>; FLYGT DX50-11 415V 1.1KW N FLT 2"CONN 10M CABLE	EACH	558.00
<NONE>; FLYGT DX50-20 415V 1.5KW N FLT 2"CONN 10M CABLE	EACH	621.55
<NONE>; FLYGT DX50-7 415V .75KW NO FLT 2"CONN 10M CABLE	EACH	455.00
<NONE>; FLYGT DXVM50-11 230V 1.1KW CFL 2"CONN 10M CABLE	EACH	558.00
<NONE>; FLYGT SX15 SG 415V 1.5KW N FLT 1.5"CONN 10M CABLE	EACH	578.56
<NONE>; FLYGT SX3 SG 415V .55KW NO FLT Y	EACH	279.00
<NONE>; FLYGT SX7 SG 415V .75KW NO FLT 1.5"CONN 10M CABLE	EACH	443.00
<NONE>; FLYGT SXM 2 230V .25KW CW FLT 1.25"CONN 10M CABLE	EACH	210.00
<NONE>; FLYGT SXM 3 230V .55KW CW FLT 1.25"CONN 10M CABLE	EACH	279.00
<NONE>; FLYGT SXM11 SG 230V 1.1KW N FLT 1.5"CONN 10M CABLE	EACH	530.00

BASIC PRICES : LABOUR, PLANT AND MATERIALS

S:MECHANICAL SERVICES - PIPE SUPPLY SYSTEMS : WORKS OF ALTERATION/SMALL WORKS/REPAIR
INDUSTR. PUMPS
FLYGT SUBMERS PUMPS

<NONE>; FLYGT SXM5 SG 230V .55KW NO FLT 1.5"CONN 10M CABLE	EACH	414.00
<NONE>; FLYGT SXM7 230V .75KW CW FLT 1.5"CONN 10M CABLE	EACH	443.00

INSULATION
MINERAL FIBRE INSULA

<NONE>; 10" ALUMINIUM BANDS	EACH	0.19
<NONE>; 114 X 30MM H&V LAG FOIL COVERED	PER METRE	61.56
<NONE>; 114X25MM CANV.COV.INSULATION FOR 4" STEEL TUBE	PER METRE	55.84
<NONE>; 12" ALUMINIUM BANDS	EACH	0.21
<NONE>; 14" ALUMINIUM BANDS	EACH	0.22
<NONE>; 16" ALUMINIUM BANDS	EACH	0.25
<NONE>; 18" ALUMINIUM BANDS	EACH	0.27
<NONE>; 21X25MM CANV.COV.INSULATION FOR 1/2"STEEL TUBE	PER METRE	19.74
<NONE>; 22 X 30MM H&V LAG FOIL COVERED	PER METRE	25.20
<NONE>; 27X25MM CANV.COV.INSULATION FOR 3/4"STEEL TUBE	PER METRE	21.63
<NONE>; 28 X 30MM H&V LAG FOIL COVERED	PER METRE	27.00
<NONE>; 34X25MM CANV.COV.INSULATION FOR 1"STEEL TUBE	PER METRE	24.23
<NONE>; 35 X 30MM H&V LAG FOIL COVERED	PER METRE	28.62
<NONE>; 42 X 30MM H&V LAG FOIL COVERED	PER METRE	31.32
<NONE>; 42X25MM CANV.COV.INSULATION FOR 1.1/4"STEELTUBE	PER METRE	26.49
<NONE>; 48 X 30MM H&V LAG FOIL COVERED	PER METRE	33.16
<NONE>; 48X25MM CANV.COV.INSULATION FOR 1.1/2"STEELTUBE	PER METRE	28.42
<NONE>; 60 X 30MM H&V LAG FOIL COVERED	PER METRE	38.10
<NONE>; 60X25MM CANV.COV.INSULATION FOR 2"STEEL TUBE	PER METRE	32.75
<NONE>; 76X25MM BR"O"H&V PIPE INSUL. FOR 2.1/2"STEELTUB	PER METRE	37.89
<NONE>; 8" ALUMINIUM BANDS	EACH	0.16
<NONE>; 89 X 30MM H&V LAG FOIL COVERED	PER METRE	47.17
<NONE>; 89X25MM CANV.COV.INSULATION FOR 3" STEEL TUBE	PER METRE	41.97

BASIC PRICES : LABOUR, PLANT AND MATERIALS

S:MECHANICAL SERVICES - PIPE SUPPLY SYSTEMS : WORKS OF ALTERATION/SMALL WORKS/REPAIR

MALLEABLES

CRANE MALL 32-150MM

<NONE>; 100MM (4") 256G/340G GALV UNION	EACH	215.13
<NONE>; 32MM (1.1/4") 161/130 BLK TEE	EACH	5.73
<NONE>; 32MM (1.1/4") 256/340 BLK UNION	EACH	10.48
<NONE>; 32MM (1.1/4") 256/340G GALV UNION	EACH	19.82
<NONE>; 40MM (1.1/2") 161/130 BLK TEE	EACH	8.80
<NONE>; 40MM (1.1/2") 256/340 BLK UNION	EACH	11.83
<NONE>; 50MM (2") 256G/340G GALV UNION	EACH	33.36

CRANE MALL 6 TO 25MM

<NONE>; 15MM (1/2") 161/130 BLK TEE	EACH	1.50
<NONE>; 15MM (1/2") 241/330 BLK UNION	EACH	4.57
<NONE>; 15MM (1/2") 241G/330G GALV UNION	EACH	8.42
<NONE>; 20MM (3/4") 163 BLK MF TEE	EACH	6.73
<NONE>; 20MM (3/4") 241/330 BLK UNION	EACH	5.36
<NONE>; 20MM (3/4") 241G/330G GALV UNION	EACH	9.85
<NONE>; 20MM (3/4") 256/340 BLK UNION	EACH	6.21
<NONE>; 25MM (1") 161/130 BLK TEE	EACH	3.13
<NONE>; 25MM (1") 241/330 BLK UNION	EACH	6.21
<NONE>; 25MM (1") 256/340 BLK UNION	EACH	5.57

GF MALL 65 TO 150MM

<NONE>; 80MM (3") 340G/256G GALV UNION	EACH	113.00

OTHER H&V PUMPS

WILO LT COMM

<NONE>; WILO SE 125TW LT COMMERCIAL PUMP BARE	EACH	409.00

OTHERS

BOSS RADIATOR VALVES

<NONE>; .25 BOSS 833S RAD.AIR COCK WHEEL OP. GT (R66/2)	EACH	1.68
<NONE>; .5 BOSS 830S SAFETY VALVE FEM 2.5-BAR SETTING (R140RY002)	EACH	5.24

CISTERNS

<NONE>; 70/50 SUPERSEAL LID FOR WIZARD - NO LONGER USED	EACH	30.99
<NONE>; BYELAW 30 KIT FOR 70/50 - NO LONGER USED	EACH	62.12

© NSR 2008-2009

BASIC PRICES : LABOUR, PLANT AND MATERIALS

S:MECHANICAL SERVICES - PIPE SUPPLY SYSTEMS : WORKS OF ALTERATION/SMALL WORKS/REPAIR
OTHERS
CISTERNS

Description	Unit	Price
<NONE>; BYELAW 30 KIT FOR K15R FILTERS CONN LID ELBOW JKT	EACH	24.83
<NONE>; BYELAW 30 KIT FOR K20R FILTERS CONN LID ELBOW JKT	EACH	27.22
<NONE>; BYELAW 30 KIT FOR K25R FILTERS CONN LID ELBOW JKT	EACH	29.29
<NONE>; BYELAW 30 KIT FOR K40R FILTERS CONN LID ELBOW JKT	EACH	39.30
<NONE>; BYELAW 30 KIT FOR K4R FILTERS CONN LID ELBOW JKT	EACH	14.41
<NONE>; BYELAW 30 KIT FOR K50R FILTERS CONN LID ELBOW JKT	EACH	43.63
<NONE>; GALVD.LID FOR 110/25 TANK 1 PIECE 24X17	EACH	18.22
<NONE>; GALVD.LID FOR 230/50 TANK 1 PIECE 29X22	EACH	25.21
<NONE>; GALVD.LID FOR 45/10 TANK 1 PIECE 18X12	EACH	16.81
<NONE>; GALVD.LID FOR 90/20 TANK 1 PIECE 24X16	EACH	18.22
<NONE>; K15R POLY F+E CISTERN RECTANGULAR KITE MARKED	EACH	33.56
<NONE>; K20R POLY F+E CISTERN RECTANGULAR KITE MARKED	EACH	33.69
<NONE>; K25R POLY F+E CISTERN RECTANGULAR KITE MARKED	EACH	44.64
<NONE>; K40R POLY F+E CISTERN RECTANGULAR KITE MARKED	EACH	77.48
<NONE>; K50R POLY F+E CISTERN RECTANGULAR KITE MARKED	EACH	83.48
<NONE>; RL15 SUPERSEAL LID FOR K15R	EACH	14.35
<NONE>; RL20 SUPERSEAL LID FOR K20R	EACH	14.83
<NONE>; RL25 SUPERSEAL LID FOR K25R	EACH	16.39
<NONE>; RL40 SUPERSEAL LID FOR K40R	EACH	26.99
<NONE>; RL50 SUPERSEAL LID FOR K50R	EACH	28.42
<NONE>; SCM110LTR/SC25GAL GALV OPEN TOP CISTERN-24X17X17	EACH	116.30
<NONE>; SCM230LTR/SC50GAL GALV OPEN TOP CISTERN-29X22X22	EACH	164.93
<NONE>; SCM45LTR/SC10GAL GALV OPEN TOP CISTERN-18X12X12	EACH	78.00
<NONE>; SCM90LTR/SC20GAL GALV OPEN TOP CISTERN-24X16X15	EACH	103.26
<NONE>; WIZARD 70/50 POLY F+E CISTERN - NO LONGER USED	EACH	67.49

CRANE J RANGE FTTGS

BASIC PRICES : LABOUR, PLANT AND MATERIALS

S:MECHANICAL SERVICES - PIPE SUPPLY SYSTEMS : WORKS OF ALTERATION/SMALL WORKS/REPAIR
OTHERS
CRANE J RANGE FTTGS

<NONE>; 15MM (1/2") J256 UNION	EACH	25.94
<NONE>; 20MM (3/4") J256 UNION	EACH	32.78

MISC ENG PRODUCTS

<NONE>; ASJ01 ARCTIC SPRAY JUMBO 500ML (41)	EACH	13.84
<NONE>; ASM01 FREEZING JKT 8-28MM (41)	EACH	1.22

MISCELLANEOUS

<NONE>; BALL FLOAT	nr	0.67
<NONE>; BALL VALVE - 0.5 IN	nr	7.64
<NONE>; BALL VALVE - 0.75 IN	nr	10.51
<NONE>; BALL VALVE ARM	nr	2.95
<NONE>; HIGH PRESSURE BALL VALVE - 0.5 IN	nr	7.64
<NONE>; HIGH PRESSURE BALL VALVE - 0.75 IN	nr	11.53
<NONE>; SIGHT GLASS WINDOW + WASHER SET	EACH	11.46

PIPED SUPPLY SYSTEMS
Piped Supply Systems

Piped Supply Systems; Copper pipes to BS EN1057 R250 with capillary joints and brass fittings - 15mm stop end	nr	2.08
Piped Supply Systems; Copper pipes to BS EN1057 R250 with capillary joints and brass fittings - 22mm stop end	nr	3.88
Piped Supply Systems; Copper pipes to BS EN1057 R250 with capillary joints and brass fittings - 28mm stop end	nr	6.92
Piped Supply Systems; Copper pipes to BS EN1057 R250 with capillary joints and brass fittings - 35mm stop end	nr	15.30
Piped Supply Systems; Copper pipes to BS EN1057 R250 with capillary joints and brass fittings - 15mm elbow	nr	0.60
Piped Supply Systems; Copper pipes to BS EN1057 R250 with capillary joints and brass fittings - 15mm equal tees	nr	1.13
Piped Supply Systems; Copper pipes to BS EN1057 R250 with capillary joints and brass fittings - 22mm elbow	nr	1.56
Piped Supply Systems; Copper pipes to BS EN1057 R250 with capillary joints and brass fittings - 22mm equal tees	nr	3.61
Piped Supply Systems; Copper pipes to BS EN1057 R250 with capillary joints and brass fittings - 28mm elbow	nr	3.08
Piped Supply Systems; Copper pipes to BS EN1057 R250 with capillary joints and brass fittings - 28mm equal tees	nr	8.55
Piped Supply Systems; Copper pipes to BS EN1057 R250 with capillary joints and brass fittings - 35mm elbow	nr	13.39
Piped Supply Systems; Copper pipes to BS EN1057 R250 with capillary joints and brass fittings - 35mm equal tees	nr	21.80

BASIC PRICES : LABOUR, PLANT AND MATERIALS

S:MECHANICAL SERVICES - PIPE SUPPLY SYSTEMS : WORKS OF ALTERATION/SMALL WORKS/REPAIR
PIPED SUPPLY SYSTEMS

Piped Supply Systems

Description	Unit	Price
Piped Supply Systems; Copper pipes to BS EN1057 R250 with capillary joints and brass fittings - 42mm elbow	nr	22.13
Piped Supply Systems; Copper pipes to BS EN1057 R250 with capillary joints and brass fittings - 42mm equal tees	nr	34.98
Piped Supply Systems; Copper pipes to BS EN1057 R250 with capillary joints and brass fittings - 42mm stop end	nr	26.34
Piped Supply Systems; Copper pipes to BS EN1057 R250 with capillary joints and brass fittings - 54mm elbow	nr	45.71
Piped Supply Systems; Copper pipes to BS EN1057 R250 with capillary joints and brass fittings - 54mm equal tees	nr	70.53
Piped Supply Systems; Copper pipes to BS EN1057 R250 with capillary joints and brass fittings - 54mm stop end	nr	36.77
Piped Supply Systems; Copper pipes to BS EN1057 R250 with compression joints - 15mm elbow	nr	2.23
Piped Supply Systems; Copper pipes to BS EN1057 R250 with compression joints - 15mm stop end	nr	2.89
Piped Supply Systems; Copper pipes to BS EN1057 R250 with compression joints - 15mm tees	nr	3.23
Piped Supply Systems; Copper pipes to BS EN1057 R250 with compression joints - 22mm elbow	nr	3.76
Piped Supply Systems; Copper pipes to BS EN1057 R250 with compression joints - 22mm stop end	nr	3.46
Piped Supply Systems; Copper pipes to BS EN1057 R250 with compression joints - 22mm tees	nr	5.28
Piped Supply Systems; Copper pipes to BS EN1057 R250 with compression joints - 28mm elbow	nr	13.49
Piped Supply Systems; Copper pipes to BS EN1057 R250 with compression joints - 28mm stop end	nr	11.23
Piped Supply Systems; Copper pipes to BS EN1057 R250 with compression joints - 28mm tees	nr	24.55
Piped Supply Systems; Copper pipes to BS EN1057 R250 with compression joints - 35mm elbow	nr	33.94
Piped Supply Systems; Copper pipes to BS EN1057 R250 with compression joints - 35mm stop end	nr	20.27
Piped Supply Systems; Copper pipes to BS EN1057 R250 with compression joints - 35mm tees	nr	44.87
Piped Supply Systems; Copper pipes to BS EN1057 R250 with compression joints - 42mm elbow	nr	47.51
Piped Supply Systems; Copper pipes to BS EN1057 R250 with compression joints - 42mm stop end	nr	33.14
Piped Supply Systems; Copper pipes to BS EN1057 R250 with compression joints - 42mm tees	nr	69.05
Piped Supply Systems; Copper pipes to BS EN1057 R250 with compression joints - 54mm elbow	nr	80.70
Piped Supply Systems; Copper pipes to BS EN1057 R250 with compression joints - 54mm stop end	nr	46.22

© NSR 2008-2009

BASIC PRICES : LABOUR, PLANT AND MATERIALS

S:MECHANICAL SERVICES - PIPE SUPPLY SYSTEMS : WORKS OF ALTERATION/SMALL WORKS/REPAIR
PIPED SUPPLY SYSTEMS

Piped Supply Systems

Piped Supply Systems; Copper pipes to BS EN1057 R250 with compression joints - 54mm tees		nr	111.12

PLASTICS

20-125MM BLUE MDPE COIL

<NONE>; 20MM MDPE BLUE TUBE 50M COIL SDR9	PER METRE	1.09
<NONE>; 25MM MDPE BLUE TUBE 50M COIL SDR11	PER METRE	1.23
<NONE>; 32MM MDPE BLUE TUBE 50M COIL SDR11	PER METRE	2.08
<NONE>; 50MM MDPE BLUE TUBE 50M COIL SDR11	PER METRE	5.06
<NONE>; 63MM MDPE BLUE TUBE 50M COIL SDR11	PER METRE	8.05
<NONE>; 90MM MDPE BLUE TUBE 50M COIL SDR11	PER METRE	15.78

BOSS MISC PLASTICS

<NONE>; 15MM CISTERMISER CTL VVE STD (LP6073A)	EACH	137.50
<NONE>; CISTERMISER CEILING MOUNTING KIT FOR IRC VALVE	EACH	25.00

DPIPE UPVC FTGS/VVES

<NONE>; 0.375 100 D/PIPE PVC SOCKET /GF219111	EACH	1.20
<NONE>; 0.50 100 D/PIPE PVC SOCKET GF219111	EACH	1.32
<NONE>; 0.75 100 D/PIPE PVC SOCKET /GF219111	EACH	1.47

GF METRIC PVCU

<NONE>; 10mm PN25 pipe	m	1.29
<NONE>; 12mm PN25 pipe	m	1.40
<NONE>; 16mm PN16 pipe	m	1.43
<NONE>; 20mm PN16 pipe	m	2.10
<NONE>; 25mm PN16 pipe	m	2.73
<NONE>; 32mm PN16 pipe	m	4.05
<NONE>; 40mm PN16 pipe	m	5.99
<NONE>; 50mm PN16 pipe	m	8.72

GF UPVC FTGS/VVES

<NONE>; 10mm 90deg Elbow	nr	1.05
<NONE>; 10mm Plain 45deg Equal Tee	nr	5.67
<NONE>; 10mm Plain Cap	nr	0.91
<NONE>; 12mm 90deg Elbow	nr	1.05
<NONE>; 12mm Plain 45deg Equal Tee	nr	5.67

BASIC PRICES : LABOUR, PLANT AND MATERIALS

S:MECHANICAL SERVICES - PIPE SUPPLY SYSTEMS : WORKS OF ALTERATION/SMALL WORKS/REPAIR
PLASTICS
GF UPVC FTGS/VVES

<NONE>; 12mm Plain Cap	nr	0.91
<NONE>; 16mm 45deg elbow	nr	2.91
<NONE>; 16mm Plain 45deg Equal Tee	nr	6.18
<NONE>; 16mm Plain Cap	nr	1.02
<NONE>; 20mm 45deg Elbow	nr	3.35
<NONE>; 20mm Plain 45deg Equal Tee	nr	8.88
<NONE>; 20mm Plain Cap	nr	1.02
<NONE>; 25mm 45deg Elbow	nr	3.55
<NONE>; 25mm Plain 45deg Equal Tee	nr	9.50
<NONE>; 25mm Plain Cap	nr	1.47
<NONE>; 32mm 45deg Elbow	nr	4.32
<NONE>; 32mm Plain 45deg Equal Tee	nr	10.68
<NONE>; 32mm Plain Cap	nr	1.63
<NONE>; 40mm 45deg Elbow	nr	6.18
<NONE>; 40mm Plain 45deg Equal Tee	nr	13.06
<NONE>; 40mm Plain Cap	nr	2.57
<NONE>; 50mm 45deg Elbow	nr	7.77
<NONE>; 50mm Plain 45deg Equal Tee	nr	16.44
<NONE>; 50mm Plain Cap	nr	4.32

M D P E YELLOW TUBE

<NONE>; 20MM MDPE YELLOW TUBE 50M COIL SDR9	PER METRE	1.45
<NONE>; 25MM MDPE YELLOW TUBE 50MCOIL SDR11	PER METRE	1.86
<NONE>; 32MM MDPE YELLOW TUBE 50M COIL SDR11	PER METRE	3.00
<NONE>; 63MM MDPE YELLOW TUBE 50M COIL SDR11	PER METRE	11.44

MDPE FITTINGS

COMPRESSION; 20MM COMPRESSION END CAP	nr	3.51
COMPRESSION; 20MM COMPRESSION JOINER	nr	3.22
COMPRESSION; 20MM COMPRESSION TEE - EQUAL	nr	5.79
COMPRESSION; 25MM COMPRESSION END CAP	nr	4.71
COMPRESSION; 25MM COMPRESSION JOINER	nr	3.39
COMPRESSION; 25MM COMPRESSION TEE - EQUAL	nr	9.06
COMPRESSION; 32MM COMPRESSION END CAP	nr	6.28
COMPRESSION; 32MM COMPRESSION JOINER	nr	8.04

BASIC PRICES : LABOUR, PLANT AND MATERIALS

S:MECHANICAL SERVICES - PIPE SUPPLY SYSTEMS : WORKS OF ALTERATION/SMALL WORKS/REPAIR
PLASTICS
MDPE FITTINGS

COMPRESSION; 32MM COMPRESSION TEE - EQUAL	nr	11.35
COMPRESSION; 50MM COMPRESSION END CAP	nr	15.17
COMPRESSION; 50MM COMPRESSION JOINER	nr	18.52
COMPRESSION; 50MM COMPRESSION TEE - EQUAL	nr	26.47
COMPRESSION; 63MM COMPRESSION END CAP	nr	21.48
COMPRESSION; 63MM COMPRESSION JOINER	nr	27.87
COMPRESSION; 63MM COMPRESSION TEE - EQUAL	nr	41.00
FUSION; 20MM FUSION COUPLER PE100	EACH	7.28
FUSION; 20MM FUSION END CAP PE100	EACH	11.88
FUSION; 20MM FUSION TEE PE100	EACH	15.10
FUSION; 25MM FUSION COUPLER PE100	EACH	7.28
FUSION; 25MM FUSION END CAP PE100	EACH	11.88
FUSION; 25MM FUSION TEE PE100	EACH	15.10
FUSION; 32MM FUSION COUPLER PE100	EACH	7.28
FUSION; 32MM FUSION END CAP PE100	EACH	11.88
FUSION; 32MM FUSION TEE PE100	EACH	17.16
FUSION; 50MM FUSION COUPLER PE100	EACH	12.08
FUSION; 50MM FUSION END CAP PE100	EACH	19.91
FUSION; 50MM FUSION TEE PE100	EACH	26.51
FUSION; 63MM FUSION COUPLER PE100	EACH	13.31
FUSION; 63MM FUSION END CAP PE100	EACH	22.02
FUSION; 63MM FUSION TEE PE100.	EACH	37.75
FUSION; 90MM FUSION COUPLER PE100	EACH	19.98
FUSION; 90MM FUSION END CAP PE100	EACH	39.12
FUSION; 90MM FUSION TEE PE100	EACH	61.42

MDPE PIPE & ANCIL

<NONE>; 63MM MDPE BLUE TUBE 6M LENGTH SDR11	PER METRE	7.98

UPVC TUBE

<NONE>; .375 UPVC TUBE CL E P/E BS3505	PER METRE	3.14
<NONE>; 0.5 UPVC TUBE.CL 7 P/E BS3505	PER METRE	6.51
<NONE>; 0.75 UPVC TUBE.CL 7 P/E BS3505	PER METRE	9.11

STRAINER+AIRPEL

BASIC PRICES : LABOUR, PLANT AND MATERIALS

S:MECHANICAL SERVICES - PIPE SUPPLY SYSTEMS : WORKS OF ALTERATION/SMALL WORKS/REPAIR
STRAINER+AIRPEL
BRONZE STRAINERS

<NONE>; 15MM 47N DZR BRASS Y STR BSPT 20MESH S/S	EACH	39.66
<NONE>; 20MM 47N DZR BRASS Y STR BSPT 20MESH S/S	EACH	52.06
<NONE>; 25MM 47N DZR BRASS Y STR BSPT 20MESH S/S	EACH	79.03
<NONE>; 32MM 47N BRONZE Y STRAINER 20MESH S/S	EACH	118.44
<NONE>; 40MM 47N BRONZE Y STRAINER 20MESH S/S	EACH	144.16
<NONE>; 50MM 47N BRONZE Y STRAINER 0.8MM PERF S/S	EACH	265.55

TA VALVES
TA BRONZE VALVES

<NONE>; 15MM MD01 METERING STATION STD FLOW AMETAL T&A	EACH	15.34
<NONE>; 20MM MD01 METERING STATION STD FLOW AMETAL T&A	EACH	21.96
<NONE>; 25MM MD01 METERING STATION STD FLOW AMETAL T&A	EACH	24.47
<NONE>; 32MM MD01 METERING STATION STD FLOW AMETAL T&A	EACH	27.93
<NONE>; 40MM MD01 METERING STATION STD FLOW AMETAL T&A	EACH	33.10
<NONE>; 50MM MD01 METERING STATION STD FLOW AMETAL T&A	EACH	43.49
<NONE>; PT100 PRESSURE TEST POINT TA	EACH	12.97

TA METRIC CI VALVES

<NONE>; 100MM MDF0 METERING STATION T&A	EACH	149.94
<NONE>; 65MM MDF0 METERING STATION T&A	EACH	108.91
<NONE>; 80MM MDF0 METERING STATION T&A	EACH	125.03

WATER HEATERS
LIFF WATER TREATMENT

<NONE>; LBC15 LIME BEATER ELECTROLYTIC SCALE INHIBITOR	EACH	48.80
<NONE>; LBC22 LIME BEATER ELETROLYTIC SCALE INHIBITOR	EACH	52.22
<NONE>; LBC28 LIME BEATER ELECTROLYTI$ SCALE INHIBITOR	EACH	300.00
<NONE>; LBC35 LIME BEATER ELECTROLYTIC SCALE INHIBITOR	EACH	360.00
<NONE>; LBC42 LIME BEATER ELECTROLYTI$ SCALE INHIBITOR	EACH	470.00

BASIC PRICES : LABOUR, PLANT AND MATERIALS

S:MECHANICAL SERVICES - PIPE SUPPLY SYSTEMS : WORKS OF ALTERATION/SMALL WORKS/REPAIR
WATER HEATERS

LIFF WATER TREATMENT

Description	Unit	Price
<NONE>; LBC54 LIME BEATER ELECTROLYTI$ SCALE INHIBITOR	EACH	480.00
<NONE>; R100N ROTOR MAG ELECTROMAGNETI PLEASE NOTE A PUMP IS REQUIRE	EACH	3,150.00
<NONE>; R125N ROTOR MAG ELECTROMAGNETI PLEASE NOTE A PUMP IS REQUIRE - DISCONTINUED	EACH	3,971.12
<NONE>; R150N ROTOR MAG ELECTROMAGNETI PLEASE NOTE A PUMP IS REQUIRE	EACH	4,270.00
<NONE>; R200N ROTOR MAG ELECTROMAGNETI PLEASE NOTE A PUMP IS REQUIRE	EACH	4,840.00
<NONE>; R300N ROTOR MAG ELECTROMAGNETI PLEASE NOTE A PUMP IS REQUIRE	EACH	5,130.00
<NONE>; R400N ROTOR MAG ELECTROMAGNETI PLEASE NOTE A PUMP IS REQUIRE	EACH	5,410.00
<NONE>; R75N ROTOR MAG ELECTROMAGNETIC PLEASE NOTE A PUMP IS REQUIRE	EACH	2,790.00
<NONE>; T1/6 SIMPLEX WATER SOFTENER TIMER CONTROL	EACH	516.00
<NONE>; T11/6 SIMPLEX WATER SOFTENER TIMER CONTROL	EACH	1,047.00
<NONE>; T16/6 SIMPLEX WATER SOFTENER TIMER CONTROL	EACH	1,578.00
<NONE>; T2/3 SIMPLEX WATER SOFTENER TIMER CONTROL	EACH	540.00
<NONE>; T3/3 SIMPLEX WATER SOFTENER TIMER CONTROL	EACH	588.00
<NONE>; T5/0 SIMPLEX WATER SOFTENER TIMER CONTROL	EACH	659.00
<NONE>; T7/5 SIMPLEX WATER SOFTENER TIMER CONTROL	EACH	735.00
<NONE>; V11/6D DUPLEX WATER SOFTENER VOLUMETRIC CONTROL	EACH	2,271.00
<NONE>; V16/6D DUPLEX WATER SOFTENER VOLUMETRIC CONTROL	EACH	2,840.00
<NONE>; V5/0D DUPLEX WATER SOFTNER VOLUMETRIC CONTROL	EACH	1,536.00
<NONE>; V7/5D DUPLEX WATER SOFTNER VOLUMETRIC CONTROL	EACH	1,734.00

WORC & NORBRO
WORC.3-PCE NON F/S

Description	Unit	Price
<NONE>; GLAND KIT REPLACEMENT	EACH	26.78

BASIC PRICES : LABOUR, PLANT AND MATERIALS

T:MECHANICAL SERVICES - MECHANICAL HEATING/COOLING REFRIGERATION SYSTEMS : WORKS OF ALTERATION/SMALL WORKS/REPAIR

AIR ELIMINATORS

BROWNALLS AIR ELIMS

<NONE>; BOSS AUTO AIR ELIMINATOR EV C/W 15MM ISO VALVE	EACH	123.12

BOSS VALVES

963-965S BALL VALVES

<NONE>; 15MM BOSS 965S BZE BALL VALVE BSP TAPER THRD.F/BORE (1300)	EACH	17.13
<NONE>; 20MM BOSS 965S BZE BALL VALVE BSP TAPER THRD.F/BORE (1300)	EACH	24.21
<NONE>; 25MM BOSS 965S BZE BALL VALVE BSP TAPER THRD.F/BORE (1300)	EACH	27.64

BOSS 348S REL VALVES

<NONE>; 15MM 348S BZE SFTY VVE 175/225	EACH	226.84
<NONE>; 50MM 348S BZE SFTY VVE 175/225	EACH	550.09

BOSS 716

<NONE>; 40X65 BOSS716ESL S/R 81-125PSI S/V BZE.EPDM DISC.5.5-8.6 BSP	EACH	619.32
<NONE>; 50X80 BOSS716ESL S/R 81-125PSI S/V BZE.EPDM DISC.5.5-8.6 BSP	EACH	902.27

COMM BOILERS

BAXI GAS BOILER WALL HUNG

<NEXT>; BAXI COMBO 80E L BOILER	EACH	761.20

NABIC VALVES

<NONE>; FIG 175, THREE WAY VENT COCK, 25MM	nr	297.12
<NONE>; FIG 175, THREE WAY VENT COCK, 50MM	nr	717.44
<NONE>; FIG 175, THREE WAY VENT COCK, 65MM	nr	907.81

POTTERTON COM BOILER

<NONE>; POTTERTON PERFORMA 24	nr	901.25

WORCESTER OIL BOILER

<NONE>; WORCESTER OIL - DANESMOOR- 100MM FLEXIBLE FLUE KIT 8MTR	nr	501.71
<NONE>; WORCESTER OIL - DANESMOOR-CF 20/25	nr	1,541.97
<NONE>; WORCESTER OIL - GREENSTAR DANESMOOR UTILITY 18-25	nr	1,993.93
<NONE>; WORCESTER OIL - STANDARD RS FLUE KIT	nr	168.41

BASIC PRICES : LABOUR, PLANT AND MATERIALS

T:MECHANICAL SERVICES - MECHANICAL HEATING/COOLING REFRIGERATION SYSTEMS : WORKS OF ALTERATION/SMALL WORKS/REPAIR
COMMERCIAL RADS

MHS CAST IRON RADIATORS & ACCESSORIES

<NONE>; CAST IRON FLOOR BRACKET	nr	12.00
<NONE>; CAST IRON RADIATOR - 635 MM HIGH X 1024 MM LONG	nr	608.00
<NONE>; CAST IRON RADIATOR - 635 MM HIGH X 524 MM LONG	nr	353.00
<NONE>; CAST IRON RADIATOR - 635 MM HIGH X 624 MM LONG	nr	404.00
<NONE>; CAST IRON RADIATOR - 635 MM HIGH X 674MM LONG	nr	430.00
<NONE>; CAST IRON RADIATOR - 635 MM HIGH X 724MM LONG	nr	455.00
<NONE>; CAST IRON RADIATOR - 635 MM HIGH X 824 MM LONG	nr	506.00
<NONE>; CAST IRON RADIATOR - 635 MM HIGH X 874 MM LONG	nr	531.00

STELRAD ELITE RADS

<NONE>; K1 700X1000 ELITE RADIATOR 4 X1/2" CONS (30 SECTIONS)	EACH	104.68
<NONE>; K1 700X1200 ELITE RADIATOR 4 X 1/2" CONS (36 SECTIONS)	EACH	168.94
<NONE>; K1 700X1800 ELITE RADIATOR 4 X 1/2" CONS (54 SECTIONS)	EACH	253.41
<NONE>; K1 700X2200 ELITE RADIATOR 4 X 1/2" CONS (66 SECTIONS)	EACH	309.73
<NONE>; K1 700X400 ELITE RADIATOR 4 X 1/2" CONS (12 SECTIONS)	EACH	36.59
<NONE>; K1 700X500 ELITE RADIATOR 4 X1/2" CONS (15 SECTIONS)	EACH	45.73
<NONE>; K1 700X600 ELITE RADIATOR 4 X 1/2" CONS (18 SECTIONS)	EACH	54.88
<NONE>; K1 700X700 ELITE RADIATOR 4 X 1/2" CONS (21 SECTIONS)	EACH	64.01
<NONE>; K1 700X800 ELITE RADIATOR 4 X 1/2" CONS (24 SECTIONS)	EACH	73.16
<NONE>; K1 700X900 ELITE RADIATOR 4 X 1/2" CONS (27 SECTIONS)	EACH	94.20
<NONE>; K2 700X1200 ELITE RADIATOR 4 X 1/2" CONS (36 SECTIONS)	EACH	307.98
<NONE>; K2 700X1800 ELITE RADIATOR 4 X 1/2" CONS (54 SECTIONS)	EACH	461.98
<NONE>; K2 700X2200 ELITE RADIATOR 4 X 1/2" CONS (66 SECTIONS)	EACH	564.64
<NONE>; K2 700X400 ELITE RADIATOR 4 X 1/2" CONS (12 SECTIONS)	EACH	76.35

BASIC PRICES : LABOUR, PLANT AND MATERIALS

T:MECHANICAL SERVICES - MECHANICAL HEATING/COOLING REFRIGERATION SYSTEMS : WORKS OF ALTERATION/SMALL WORKS/REPAIR

COMMERCIAL RADS

STELRAD ELITE RADS

<NONE>; K2 700X500 ELITE RADIATOR 4 X 1/2" CONS (15 SECTIONS)	EACH	95.41
<NONE>; K2 700X600 ELITE RADIATOR 4 X 1/2" CONS (18 SECTIONS)	EACH	114.49
<NONE>; K2 700X700 ELITE RADIATOR 4 X 1/2" CONS (21 SECTIONS)	EACH	133.59
<NONE>; K2 700X800 ELITE RADIATOR 4 X 1/2" CONS (24 SECTIONS)	EACH	152.68
<NONE>; K2 700X900 ELITE RADIATOR 4 X 1/2" CONS (27 SECTIONS)	EACH	171.75

COPPER FTGS+VVE

COMP: CONEX A

<NONE>; 22MM FIG 301 STRAIGHT COUPLER	EACH	3.15
<NONE>; 28MM X 1 FIG 302 ST COUPLER	EACH	5.38

COMP: CONEX MAIN

<NONE>; 35MM FIG 301 STRAIGHT COUPLER	EACH	25.76

DELCOP MAIN

<NONE>; 15MM 733 UNION COUPLING	EACH	12.21
<NONE>; 28MM 601 STRT COUPLING	EACH	1.77
<NONE>; 6X6MM 601 STR COUPLING COPPERXCOPPER DELCOP	EACH	1.59
<NONE>; 8MM 601 STRT COUPLING	EACH	1.59

SOLDER RING: MSR

<NONE>; 15MM COUPLER MSR1 (100/400)	EACH	0.29
<NONE>; 28MM COUPLER MSR1 (25/200)	EACH	1.75

SOLDER RING:YORK GEN

<NONE>; 10MM YP1 COUPLING	EACH	0.91
<NONE>; 6MM YP1 STRAIGHT COUPLING	EACH	1.75

SOLDER RING:YORK YPS

<NONE>; 15MM YPS1 COUPLING	EACH	0.33
<NONE>; 22MM YPS1 COUPLING	EACH	0.85

COPPER TUBE

TUBE: WTF: 15-28 T X

<NONE>; 15X0.7X6M CU-TUBE (TX) EN1057 R250 FORMERLY BS2871 TX	PER METRE	6.72

BASIC PRICES : LABOUR, PLANT AND MATERIALS

T:MECHANICAL SERVICES - MECHANICAL HEATING/COOLING REFRIGERATION SYSTEMS : WORKS OF ALTERATION/SMALL WORKS/REPAIR

COPPER TUBE

TUBE: WTF: 15-28 T X

<NONE>; 28X0.9X6M CU-TUBE (TX) EN1057 R250 FORMERLY BS2871 TX	PER METRE	16.95

TUBE: WTF: 6-10 T W

<NONE>; 10X0.7X10M CU-TUBE (TW) EN1057 R220 FORMERLY BS2871 TW	PER METRE	6.25
<NONE>; 6X0.6X10M CU-TUBE (TW) EN1057 R220 FORMERLY BS2871 TW	PER METRE	3.55
<NONE>; 8X0.6X10M CU-TUBE (TW) EN1057 R220 FORMERLY BS2871 TW	PER METRE	4.54

TUBE: WTF:REMAINDER

<NONE>; 22X0.9X3M CP CU-TUBE (TX) EN1057 R250 FORMERLY BS2871 TX	PER METRE	22.65

TUBE: YCT: 15-28 T.X

<NONE>; 15X0.7X2M CU-TUBE (TX) EN1057 R250 FORMERLY BS2871 TX	PER METRE	6.72
<NONE>; 15X0.7X6M CU-TUBE (TX) EN1057 R250 FORMERLY BS2871 TX	PER METRE	6.72
<NONE>; 22X0.9X6M CU-TUBE (TX) EN1057 R250 FORMERLY BS2871 TX	PER METRE	13.44
<NONE>; 28X0.9X6M CU-TUBE (TX) EN1057 R250 FORMERLY BS2871 TX	PER METRE	16.95

TUBE: YCT: 6-1- T W

<NONE>; 6X0.6X25M CU-TUBE (TW) EN1057 R220 FORMERLY BS2871 TW	PER METRE	3.55

ELECTRIC MOTORS

ELECTRIC MOTOR SPARES & ACCESSORIES

SPARES/REPLACEMENT PARTS; FAN ELEC12506_D90FAN - NO LONGER USED	Item	18.36

ELECTRIC MOTORS - THREE PHASE

3 PHASE 50HZ ALUMINIUM. FOOT & FLANGE MOUNTING. FRAMES 63-250. 0.18KW-55KW; D63 0.18KW FOOT MTG 3PHASE 50HZ ALUM MOTOR D63-.18KW1500FM - NO LONGER USED	Item	96.64

FLUE & CHIMNEY

MISC FLUE

<NONE>; 125MM SW 970MM FLUE LGTH (003) SELDEK SW	EACH	24.21
<NONE>; 125MM SW LOCKING BAND (086) SELKIRK SW	EACH	10.97

BASIC PRICES : LABOUR, PLANT AND MATERIALS

T:MECHANICAL SERVICES - MECHANICAL HEATING/COOLING REFRIGERATION SYSTEMS : WORKS OF ALTERATION/SMALL WORKS/REPAIR

GAUGES

GAUGE SYPHONS

<NONE>; 15MM 401S MS U SYPHON	EACH	26.43

GAUGES & THERMOMETER

<NONE>; 100MM 407DM P/GAUGE 10BAR/150 BRASS CASE	EACH	85.74
<NONE>; 200MM STR THERMOMETER	EACH	41.15

GENERAL

Sundries

Materials; Unit material rate	BASE	1.35

GRUNDFOS PUMPS

GRUNDFOS UPE PUMPS

<NONE>; UPED32-120F 1PH TWIN USE 83530169	EACH	1,519.00
<NONE>; UPED65-60F 1PH TWIN USE 83530221	EACH	2,234.00
<NONE>; UPED80-120F 3PH TWIN	EACH	3,795.00

HATTERSLEY VVES

HATTS BRONZE VALVES

<NONE>; 15MM(1/2") 2086TRV BZE D/FLO POL CHROME ANGLE PATT HATTS	EACH	42.23
<NONE>; 15MM(1/2") 3180 BZE D/FLO VV RAD - MATT CH ANGLE PATT HATT	EACH	18.14
<NONE>; 22MM 3180 BZE D/FLO VVE RAD - MATT CH ANGLE PATT HATTS	EACH	23.06
<NONE>; 22MM(3/4") 2086TRV BZE D/FLO POL CHROME ANGLE PATT HATTS	EACH	54.16
<NONE>; 25MM(1") 2380LS BZE D/FLO VVE RAD - MATT CH ANGLE PATT HATTS	EACH	39.27

HEAT EMITTERS

BOSS PERIMETER HTG

<NONE>; 165HX90D TOP OUTLET WALL BRKT FIRST FIX ITEM	EACH	17.85
<NONE>; 22X1080MM 76X76MM N ELEMENT	EACH	55.65
<NONE>; 22X2080MM 108X108MM W ELEMENT	EACH	154.35
<NONE>; 28X1080MM 76X76MM N ELEMENT	EACH	61.95
<NONE>; 28X2080MM 108X108MM W ELEMENT	EACH	158.55
<NONE>; 28X2080MM 76X76MM N ELEMENT	EACH	119.70

BASIC PRICES : LABOUR, PLANT AND MATERIALS

T:MECHANICAL SERVICES - MECHANICAL HEATING/COOLING REFRIGERATION SYSTEMS : WORKS OF ALTERATION/SMALL WORKS/REPAIR

HEAT EMITTERS

BOSS PERIMETER HTG

<NONE>; CE5/15 .5M ELEMENT-15MM	EACH	8.11

COPPERAD FAN CONVECT

<NONE>; 2005S/T1 FAN CONVECTOR	EACH	875.01
<NONE>; 2009S/T1 FAN CONVECTOR	EACH	998.58
<NONE>; 2015S/T1 FAN CONVECTOR	EACH	1,237.63
<NONE>; 2104/CS/LTC/K FAN CONVECTOR	EACH	807.25
<NONE>; 2112/CS/LTC/K FAN CONVECTOR	EACH	1,109.00
<NONE>; 2115/CS/LTC/K FAN CONVECTOR	EACH	1,361.79
<NONE>; SS04/LTC FAN CONVECTOR	EACH	838.42
<NONE>; SS09/LTC FAN CONVECTOR	EACH	1,024.06
<NONE>; SS15/LTC FAN CONVECTOR	EACH	1,263.58

COPPERAD NATURAL CON

<NONE>; 15040 COPPERAD CONVECTOR FULL LENGTH GRILLE NO DAMPER	EACH	270.00
<NONE>; 15080 COPPERAD CONVECTOR FULL LENGTH GRILLE NO DAMPER	EACH	305.00
<NONE>; 15120 COPPERAD CONVECTOR FULL LENGTH GRILLE NO DAMPER	EACH	350.00
<NONE>; 15140 COPPERAD CONVECTOR FULL LENGTH GRILLE NO DAMPER	EACH	400.00

COPPERAD SPARES

<NONE>; 4 POS SWITCH NO WIRE	EACH	25.00

COPPERAD UNIT HEATER

<NONE>; A14/WG COPPERAD UNIT HEATER 3 SPIDER ARM MOTOR	EACH	807.48
<NONE>; DW27/D2 COPPERAD UNIT HEATER 3 SPIDER ARM MOTOR	EACH	1,419.46

VA AIR CURTAIN

<NONE>; VA 100CM, 6KW 415V AIR CURTAIN REF:431822	nr	736.04
<NONE>; VA 150CM, 9KW 415V AIR CURTAIN REF:431823	nr	913.79
<NONE>; VA 200CM, 12KW 415V AIR CURTAIN REF:431824	nr	1,002.38

HEATING CONTROL

ALCON INCENTIVE ITEM

<NONE>; GB2C 1/4 GAS VVE FAST OPEN240V WAS GB2A	EACH	64.83

BASIC PRICES : LABOUR, PLANT AND MATERIALS

T:MECHANICAL SERVICES - MECHANICAL HEATING/COOLING REFRIGERATION SYSTEMS : WORKS OF ALTERATION/SMALL WORKS/REPAIR
HEATING CONTROL

BLACK TEKNIGAS

<NONE>; 70003 SERIES UNIVERSAL THERMOCOUPLE 900MM	EACH	5.20
<NONE>; ME39SS BLACK 6" ELECTRODE	EACH	27.90
<NONE>; SOLENOID GAS VALVE 230V 3/4"	EACH	279.33

DANFOSS TRV

<NONE>; 05 RA-G15 ANG VALVE BODY (13G0123) NEW 13G3383	EACH	31.63
<NONE>; 075 RA-G20 ANGLE VALV BODY (013G0125) NEW 013G3385	EACH	34.12
<NONE>; 1.00 RA-G25 ANG VALV BODY (013G0127) NEW 013G3387	EACH	45.24
<NONE>; RA GLAND SEAL 13GO290	EACH	6.86
<NONE>; RA SENSOR RTA 2M CAP 13G5062 /13G2062	EACH	61.92
<NONE>; RA STD SENSOR REMOTE 13G291200	EACH	28.75
<NONE>; RA T/PROOFSENSOR REM.13G292200	EACH	33.87

DRAYTON TRV

<NONE>; 0705150 15MM ANGLE TRV4	EACH	23.50
<NONE>; 0705154 .75" ANGLE TRV4	EACH	37.27
<NONE>; 0715621 BODY .5" ANGLE SP TRV4 (WAS 0704621)	EACH	33.94
<NONE>; 0715622 BODY .75" ANG SP TRV4 (WAS 0704622)	EACH	54.65
<NONE>; 0715623 BODY ONLY 1" ANGLE SP TRV4	EACH	49.43
<NONE>; 0725006 TRV4 INTEG HEAD ONLY (WAS 0794006)	EACH	20.25
<NONE>; TRV4 INTEGRAL HEAD FOR TRV3	EACH	26.19
<NONE>; TRV4 REMOTE HEAD FOR TRV3 TRV4REMCONHEAD/0794013	EACH	30.22

HEATING SPARES

<NONE>; 30H8914 H TYPE HOLLOW NOZZLE	EACH	6.26

HNYWELL HVAC CONTROL

<NONE>; ML7984A4009 ACTUATOR 24V	EACH	165.00

HNYWELL M/SEAT VVES

<NONE>; 15 - 50mm DIAMETER LINE SIZE MOTOR ASSEMBLY FOR CONTROL VALVE P.C SUM £200	nr	200.00
<NONE>; 65 - 100mm DIAMETER LINE SIZE MOTOR ASSEMBLY FOR CONTROL VALVE P.C. SUM £300	nr	300.00
<NONE>; M77061E1012 ACTUATOR	EACH	240.00
<NONE>; V5011S1070 1.1/4BSP 2PORT VVE WAS V5011A8218	EACH	110.00

BASIC PRICES : LABOUR, PLANT AND MATERIALS

T:MECHANICAL SERVICES - MECHANICAL HEATING/COOLING REFRIGERATION SYSTEMS : WORKS OF ALTERATION/SMALL WORKS/REPAIR
HEATING CONTROL
HNYWELL M/SEAT VVES

<NONE>; V5011S1096 2BSP 2 PORTVVE WAS V5011A8234 (OBSOLETE)	EACH	174.00
<NONE>; V5013R1057 3/4BSP 3PORT VVE WAS V5013A8026	EACH	78.00
<NONE>; V5013R1081 1.1/2BSP 3PORT VVE WAS V5013A8059	EACH	120.00
<NONE>; V5015A1151 4F 3PORT VVE	EACH	660.00
<NONE>; V5049A1599 2PORT VVE	EACH	892.00
<NONE>; V5328A1104 2.1/2F 2PT VALVE $	EACH	394.00
<NONE>; V5328A1112 3F 2PT VVE	EACH	582.00
<NONE>; V5329C1075 2.1/2F3PORTVVE	EACH	316.00
<NONE>; V5329C1083-1 3F3PORTVVE	EACH	408.00

JEAVONS

<NONE>; 1/2" J78R REGULATOR 15-23 MBAR	EACH	23.60

LANDON KINGSWAY

<NONE>; 10/0045/01 CABLE CONN	EACH	1.91
<NONE>; 10/0157/08 9M (30FT) SS CABLE	EACH	33.93
<NONE>; 10/0328/01 TENSION SPRING TO SUIT 15MM-32MM VALVES	EACH	4.33
<NONE>; 10/0329/02 BRASS SLOTTED LINK	EACH	15.77
<NONE>; 10/0331/02 A LEVER TO SUIT 1/2" TO 1" VALVES	EACH	78.41
<NONE>; 10/0335/01 SMALL BAG OF FTGS	EACH	79.40
<NONE>; 10/9003/01 MAN QUICK RELEASE	EACH	116.49
<NONE>; 10/9004/01 SPCO MERCURY SWITCH	EACH	146.33
<NONE>; 10/9028/01 FUSIBLE LINK 160DEG	EACH	6.95
<NONE>; 10/9036/05 PULLEY BSP TO BSS	EACH	16.27
<NONE>; 10/9038/02 PANIC BUTTON	EACH	17.64

TIMESWITCHES

<NONE>; 16022 COMPACT QUARTZ TIME SWITHC SINGLE CHANNEL 24HR	nr	30.57
<NONE>; 16721 COMPACT QUARTZ TIME SWITCH SINGLE CHANNEL 7DAY	nr	38.16
<NONE>; E854 ELECTRONIC TIMER ROUND PATTERN 3 PIN 7DAY	nr	171.57
<NONE>; E855 ELECTRONIC TIMER ROUND PATTERN 4 PIN 7DAY	nr	174.45
<NONE>; Q554 FORM 2 TIME SWITCH ROUND PATTERN 3 PIN 24HR	nr	125.38

BASIC PRICES : LABOUR, PLANT AND MATERIALS

T:MECHANICAL SERVICES - MECHANICAL HEATING/COOLING REFRIGERATION SYSTEMS : WORKS OF ALTERATION/SMALL WORKS/REPAIR

HEATING CONTROL
TIMESWITCHES

<NONE>; Q555 FORM 2 TIME SWITCH ROUND PATTERN 4 PIN 24HR	nr	130.29

HOT AND COLD WATER SERVICES
Water heaters

Water heaters; Electric water heater thermal cut out	nr	10.35
Water heaters; Electric water heater thermostat TST18	nr	8.25
Water heaters; Water heater Std spout and valve	nr	38.63
Water heaters; Water heater, immersion cistern, 125 litre capacity Megaflow MFCL125HE	nr	1,123.45
Water heaters; Water heater, immersion cistern, 70 litre capacity Megaflow MFCL70HE	nr	983.00

INSULATION
FLEX FOAM INSULATION

<NONE>; 13027 AF/ARMAFLEX SLIT 13MMX2M CL.1 FOR .75" STEEL TUBE	EACH	6.78
<NONE>; 19042 AF/ARMAFLEX SLIT 33.7/35x19MMX2M CL.1 FOR 1.25" STEEL TUBE	EACH	19.58

OTHER H&V PUMPS
MYSON/SMC

<NONE>; SMC SE20B THREADED BARE BRONZE LIGHT COMMERCIAL CIRCULAR	EACH	169.48

WILO LT COMM

<NONE>; WILO SE 200TW LTCOM PUMP BARE	EACH	727.00
<NONE>; WILO SE 125TW LT COMMERCIAL PUMP BARE	EACH	409.00
<NONE>; WILO TOP SD40/10 1PH TWIN HEAD PN6	EACH	1,183.00

WILO PUMPS

<NONE>; WILO IPN100/180-2.2/4 3PH IN-LINE FLGD PUMP	EACH	1,105.84
<NONE>; WILO IPN100/200-3/4 3PH IN-LINE FLANGED PUMP C/W FLGS	EACH	1,142.87
<NONE>; WILO IPN125/200-4/4 3PH IN-LINE FLGD PUMP	EACH	1,298.96
<NONE>; WILO IPN40/125-0.55/4 1PH IN-LINE FLGD PUMP	EACH	548.95
<NONE>; WILO IPN50/250-2.2/4 3PH IN-LINE FLGD PUMP	EACH	1,001.34

OTHERS
BOSS RADIATOR VALVES

BASIC PRICES : LABOUR, PLANT AND MATERIALS

T:MECHANICAL SERVICES - MECHANICAL HEATING/COOLING REFRIGERATION SYSTEMS : WORKS OF ALTERATION/SMALL WORKS/REPAIR
OTHERS
BOSS RADIATOR VALVES

<NONE>; .25 BOSS 833S RAD.AIR COCK WHEEL OP. GT (R66/2)	EACH	1.68
<NONE>; .5 BOSS 831S SAFETY VALVE MALE 2.5 BAR SETTING (R140SY002)	EACH	5.49
<NONE>; .75 BOSS 832S DIFF.BY-PASS VVE (R147Y004)	EACH	24.50
<NONE>; 15MM BOSS 829S RAD.VVE POLISHED CHROME W/H	EACH	4.81
<NONE>; 15MM BOSS TRV ANGLE PATT	EACH	13.50

MISCELLANEOUS

<NONE>; COMPRESSOR HERMTICALLY SEALED$ REF NO AJ5515E CSR	EACH	410.00

PLASTICS
DURAPIPE PVCU PIPE

<NONE>; .5"PVCU TUBE CL E	EACH	3.67

POWERFLEX
METALLIC BELLOWS

<NONE>; 100NB/28 MVT SWA WLD END EX JT	EACH	238.00
<NONE>; 15NB/25 MVT SWA WLD END EX JT	EACH	147.00
<NONE>; 20NB/25MVT SWA WLD END EX JT	EACH	139.00
<NONE>; 25NB/25 MVT SWA WLD END EX JT	EACH	140.00
<NONE>; 32NB/25 MVT SWA WLD END EX JT	EACH	164.00
<NONE>; 40NB/25 MVT SWA WLD END EX JT	EACH	169.00
<NONE>; 50NB/16 MVT SWA WLD END EX JT	EACH	142.00
<NONE>; 65NB/20 MVT SWA WLD END EX JT	EACH	164.00
<NONE>; 80NB/23 MVT SWA WLD END EX JT	EACH	192.00

RADIANT HEATING
BOSS RADLINE &PANELS

<NONE>; RLA 12 BOSS RAD-LINE 1 TUBEX2M INSULATED BACK	EACH	122.00
<NONE>; RLA 14 BOSS RAD-LINE 1 TUBEX4M INSULATED BACK	EACH	144.00
<NONE>; RLA 16 BOSS RAD-LINE 1 TUBEX6M INSULATED BACK	EACH	201.00
<NONE>; RLA 22 BOSS RAD-LINE 2 TUBEX2M INSULATED BACK	EACH	132.00

BASIC PRICES : LABOUR, PLANT AND MATERIALS

T:MECHANICAL SERVICES - MECHANICAL HEATING/COOLING REFRIGERATION SYSTEMS : WORKS OF ALTERATION/SMALL WORKS/REPAIR
RADIANT HEATING
BOSS RADLINE &PANELS

<NONE>; RLA 24 BOSS RAD-LINE 2 TUBEX4M INSULATED BACK	EACH	267.00
<NONE>; RLA 26 BOSS RAD-LINE 2 TUBEX6M INSULATED BACK	EACH	382.00
<NONE>; RLA 32 BOSS RAD-LINE 3 TUBEX2M INSULATED BACK	EACH	193.00
<NONE>; RLA 34 BOSS RAD-LINE 3 TUBEX4M INSULATED BACK	EACH	391.00
<NONE>; RLA 36 BOSS RAD-LINE 3 TUBEX6M INSULATED BACK	EACH	562.00

SENIORFLEXONICS
BOSS FAN COIL HOSES

<NONE>; BOSS FCH 15-300-AB 25092130 EPDM 304SS BRAID C/W WASHER	EACH	6.13
<NONE>; BOSS FCH 15-450-AB 25092145 EPDM 304SS BRAID C/W WASHER	EACH	7.01
<NONE>; BOSS FCH 15-600-AB 25092160 EPDM 304SS BRAID C/W WASHER	EACH	7.91
<NONE>; BOSS FCH 20-300-AB 25112730 EPDM 304SS BRAID C/W WASHER	EACH	8.18
<NONE>; BOSS FCH 20-600-AB 25112760 EPDM 304SS BRAID C/W WASHER	EACH	11.03

METAL EXP JTS SFLEX

<NONE>; 15MM/13MVT RCA PERIM'COMP	EACH	66.00
<NONE>; 22MM/25MVT RCA PERIM'COMP	EACH	89.00
<NONE>; 28MM/25MVT RCA PERIM'COMP	EACH	88.00
<NONE>; 35MM/25MVT RCA PERIM'COMP	EACH	88.00

RUBBER BELLOWS SFLEX

<NONE>; .75"BSP FTU RUBBER EXP JT	EACH	21.00
<NONE>; 1"BSP FTU RUBBER EXP JT	EACH	26.00
<NONE>; 1.25"BSP FTU RUBBER EXP JT	EACH	30.00
<NONE>; 100NB EFLEX RSPOT 4504/10/16JT	EACH	167.00
<NONE>; 2."BSP FTU RUBBER EXP JT	EACH	48.00
<NONE>; 3." BSP FTU RUBBER EXP JT	EACH	93.00

SFLEX BELLOWS

<NONE>; 100NB SFLEX EPDM SWIVEL PN16 TIE-BAR KIT IF REQD 48030158	EACH	160.00

BASIC PRICES : LABOUR, PLANT AND MATERIALS

T:MECHANICAL SERVICES - MECHANICAL HEATING/COOLING REFRIGERATION SYSTEMS : WORKS OF ALTERATION/SMALL WORKS/REPAIR

WATER HEATERS

ELECT STORAGE W/HTRS

Description	Unit	Price
<NONE>; STEIBEL-30LTR ELECTRIC WATER HTR SNF25	EACH	355.95
<NONE>; STEIBEL-50LTR ELECTRIC WATER HTR SNF50	EACH	426.30
<NONE>; STEIBEL-75LTR ELECTRIC WATER HTRSNF75	EACH	581.70
<NONE>; ZIP -150LTR ELECTRIC WTR HTR AF1150	EACH	610.00

HEATRAE SADIA WATER HEATERS

Description	Unit	Price
<NONE>; EXPRESS 15 3KW, 15 LTR WATER HEATER	EACH	351.52
<NONE>; STREAMLINE 10/3 C/W SPOUT &VVE 10 LTR 3KW	EACH	276.64

TRITON WATER HEATERS

Description	Unit	Price
<NONE>; INSTAFLOW 10 UNVENTED WTR HTR 10LTR 240V C/W RELIEF VALVE - DISCONTINUED	EACH	178.75

WATER HEATING

PLUMBING FITTINGS

Description	Unit	Price
COPPER TUBE AND FITTINGS; CHROME COMP. SERVICE VALVE 15MM	Item	5.23
COPPER TUBE AND FITTINGS; CHROME TUBE (2MTR) 15MM	Item	12.21

WATER HEATER SPARES

Description	Unit	Price
ZIP HYDROBOIL SPARES; ELEMENT + GASKET 3KW 240V 10 TO 25LITRES ELEC18376_SP6603A	Item	43.43

BASIC PRICES : LABOUR, PLANT AND MATERIALS

U:MECHANICAL SERVICES - VENTILATION/AIR CONDITIONING SYSTEMS : WORKS OF ALTERATION/SMALL WORKS/REPAIR
AIR CON+VENT
BOSS FAN COIL UNITS

<NONE>; HC2230/31 BOSS FAN COIL UNIT	EACH	622.00
<NONE>; HC2240/41 BOSS FAN COIL UNIT	EACH	666.00
<NONE>; HC3130/31 BOSS FAN COIL UNIT	EACH	686.00
<NONE>; HC3140/41 BOSS FAN COIL UNIT	EACH	732.00
<NONE>; HC4530/31 BOSS FAN COIL UNIT	EACH	756.00
<NONE>; HC4540/41 BOSS FAN COIL UNIT	EACH	808.00
<NONE>; HC6030/31 BOSS FAN COIL UNIT	EACH	908.00
<NONE>; HC6040/41 BOSS FAN COIL UNIT	EACH	958.00
<NONE>; HC7630/31 BOSS FAN COIL UNIT	EACH	1,024.00
<NONE>; HC7640/41 BOSS FAN COIL UNIT	EACH	1,092.00

DAIKIN

NONE; CEILING MOUNTED COOLING ONLY INVERTER TYPE, CASSETTE UNIT NOMINAL COOLING 3.4KW REF: FCQ35C/RKS35E	nr	1,015.00
NONE; CEILING MOUNTED COOLING ONLY INVERTER TYPE, CASSETTE UNIT NOMINAL COOLING 5.0KW REF: FCQ50C/RKS50F	nr	1,065.00
NONE; CEILING MOUNTED COOLING ONLY INVERTER TYPE, CASSETTE UNIT NOMINAL COOLING 5.0KW REF: FCQ60C/RKS60F	nr	1,145.00
NONE; CEILING MOUNTED HEAT PUMP INVERTER TYPE, CASSETTE UNIT NOMINAL COOLING 10KW REF: FCQ100C/RZQS100BV3	nr	1,770.00
NONE; CEILING MOUNTED HEAT PUMP INVERTER TYPE, CASSETTE UNIT NOMINAL COOLING 3.4KW REF: FCQ35C/RXS35E	nr	1,015.00
NONE; CEILING MOUNTED HEAT PUMP INVERTER TYPE, CASSETTE UNIT NOMINAL COOLING 5.7KW REF: FCQ60C/RXS60F	nr	1,225.00
NONE; COMPRESSOR 3 - 6KW	nr	680.00
NONE; COMPRESSOR OVER 6KW	nr	950.00
NONE; COMPRESSOR UP TO 3KW	nr	355.00
NONE; CONDENSING UNIT BRACKET - SIZE 100	nr	40.00
NONE; CONDENSING UNIT BRACKET - UP TO SIZE 71	nr	33.00
NONE; WALL MOUNTED COOLING ONLY INVERTER TYPE, NOMINAL COOLING 4.3KW REF: FTKS35DW/RKS35E	nr	540.00
NONE; WALL MOUNTED COOLING ONLY INVERTER TYPE, NOMINAL COOLING 6.0KW REF: FTKS60F/RKS60F	nr	775.00
NONE; WALL MOUNTED COOLING ONLY INVERTER TYPE, NOMINAL COOLING 7.1KW REF: FTKS71F/RKS71F	nr	1,070.00

BASIC PRICES : LABOUR, PLANT AND MATERIALS

U:MECHANICAL SERVICES - VENTILATION/AIR CONDITIONING SYSTEMS : WORKS OF ALTERATION/SMALL WORKS/REPAIR

AIR CON+VENT

DAIKIN

NONE; WALL MOUNTED HEAT PUMP INVERTER TYPE, NOMINAL COOLING 10.0KW REF: FAQ100B/RZQS100CV1	nr	1,750.00
NONE; WALL MOUNTED HEAT PUMP INVERTER TYPE, NOMINAL COOLING 3.40KW REF: FTXS35DL/RXS35E	nr	595.00
NONE; WALL MOUNTED HEAT PUMP INVERTER TYPE, NOMINAL COOLING 6.0KW REF: FTXS60F/RXS60F	nr	855.00
NONE; WALL MOUNTED HEAT PUMP INVERTER TYPE, NOMINAL COOLING 7.1KW REF: FTXS71F/RXS71F	nr	1,150.00
NONE; WIRED REMOTE CONTROLLER FOR FTXS/FTKS	nr	50.00
NONE; WIRELESS REMOTE CONTROL - FAQ100 HEAT PUMP	nr	99.00
NONE; WIRELESS REMOTE CONTROL FOR COOLING ONLY REF:BRC7F533F	nr	99.00
NONE; WIRELESS REMOTE CONTROL FOR HEAT PUMP REF:BRC7F532F	nr	79.00

SUNDRIES

NONE; 1.90LT PUMP - LITTLE GIANT VCMA205	nr	95.00
NONE; R22 REFRIGERANT	kg	21.00
NONE; R410A REFRIGERANT	kg	30.00

VENTILATION PRODUCTS

<NONE>; DUCT CIRC PRESSED BEND - 160 mm diameter	EACH	21.01
<NONE>; DUCT CIRC PRESSED BEND - 180 mm diameter	EACH	32.32
<NONE>; DUCT CIRC SEGMENTED BEND - 200 mm diameter	EACH	23.82
<NONE>; DUCT CIRC SEGMENTED BEND - 250 mm diameter	EACH	31.40
<NONE>; DUCT CIRC SEGMENTED BEND - 315 mm diameter	EACH	38.98
<NONE>; DUCT CIRC SEGMENTED BEND - 350 mm diameter	EACH	43.31
<NONE>; DUCT CIRC SEGMENTED BEND - 400 mm diameter	EACH	61.72
<NONE>; DUCT CIRC SEGMENTED BEND - 450 mm diameter	EACH	72.55
<NONE>; DUCT CIRC SEGMENTED BEND - 500 mm diameter	EACH	83.37
<NONE>; DUCT CIRC TAPER BRANCH - 160 mm diameter	EACH	34.25
<NONE>; DUCT CIRC TAPER BRANCH - 180 mm diameter	EACH	34.25
<NONE>; DUCT CIRC TAPER BRANCH - 200 mm diameter	EACH	46.83
<NONE>; DUCT CIRC TAPER BRANCH - 250 mm diameter	EACH	46.83
<NONE>; DUCT CIRC TAPER BRANCH - 315 mm diameter	EACH	64.26
<NONE>; DUCT CIRC TAPER BRANCH - 350 mm diameter	EACH	64.26
<NONE>; DUCT CIRC TAPER BRANCH - 400 mm diameter	EACH	84.64
<NONE>; DUCT CIRC TAPER BRANCH - 450 mm diameter	EACH	84.64

BASIC PRICES : LABOUR, PLANT AND MATERIALS

U:MECHANICAL SERVICES - VENTILATION/AIR CONDITIONING SYSTEMS : WORKS OF ALTERATION/SMALL WORKS/REPAIR
AIR CON+VENT
VENTILATION PRODUCTS

Description	Unit	Price
<NONE>; DUCT CIRC TAPER BRANCH - 500 mm diameter	EACH	84.64
<NONE>; DUCT RECT BEND - sum of two sides 200 mm	m	45.06
<NONE>; DUCT RECT BEND - sum of two sides 250 mm	m	50.03
<NONE>; DUCT RECT BEND - sum of two sides 300 mm	m	50.26
<NONE>; DUCT RECT BEND - sum of two sides 350 mm	m	50.52
<NONE>; DUCT RECT BEND - sum of two sides 400 mm	m	50.75
<NONE>; DUCT RECT BEND - sum of two sides 450 mm	m	52.80
<NONE>; DUCT RECT BEND - sum of two sides 500 mm	m	53.89
<NONE>; DUCT RECT BEND - sum of two sides 550 mm	m	54.43
<NONE>; DUCT RECT BEND - sum of two sides 600 mm	m	54.98
<NONE>; DUCT RECT BEND - sum of two sides 650 mm	m	56.07
<NONE>; DUCT RECT BEND - sum of two sides 700 mm	m	58.68
<NONE>; DUCT RECT BEND - sum of two sides 750 mm	m	58.96
<NONE>; DUCT RECT BEND - sum of two sides 800 mm	m	65.36
<NONE>; DUCT RECT BRANCH - sum of two sides 200 mm	m	42.46
<NONE>; DUCT RECT BRANCH - sum of two sides 250 mm	m	42.65
<NONE>; DUCT RECT BRANCH - sum of two sides 300 mm	m	42.86
<NONE>; DUCT RECT BRANCH - sum of two sides 350 mm	m	42.99
<NONE>; DUCT RECT BRANCH - sum of two sides 400 mm	m	43.20
<NONE>; DUCT RECT BRANCH - sum of two sides 450 mm	m	43.41
<NONE>; DUCT RECT BRANCH - sum of two sides 500 mm	m	45.17
<NONE>; DUCT RECT BRANCH - sum of two sides 550 mm	m	45.39
<NONE>; DUCT RECT BRANCH - sum of two sides 600 mm	m	45.60
<NONE>; DUCT RECT BRANCH - sum of two sides 650 mm	m	45.73
<NONE>; DUCT RECT BRANCH - sum of two sides 700 mm	m	45.93
<NONE>; DUCT RECT BRANCH - sum of two sides 750 mm	m	46.16
<NONE>; DUCT RECT BRANCH - sum of two sides 800 mm	m	46.26
<NONE>; STRAIGHT CIRC DUCTWORK - 160 mm diameter	m	9.55
<NONE>; STRAIGHT CIRC DUCTWORK - 160 mm diameter clips	EACH	3.46
<NONE>; STRAIGHT CIRC DUCTWORK - 180 mm diameter	m	10.30
<NONE>; STRAIGHT CIRC DUCTWORK - 180 mm diameter clips	EACH	5.41
<NONE>; STRAIGHT CIRC DUCTWORK - 200 mm diameter	m	11.16

BASIC PRICES : LABOUR, PLANT AND MATERIALS

U:MECHANICAL SERVICES - VENTILATION/AIR CONDITIONING SYSTEMS : WORKS OF ALTERATION/SMALL WORKS/REPAIR

AIR CON+VENT

VENTILATION PRODUCTS

<NONE>; STRAIGHT CIRC DUCTWORK - 200 mm diameter clips	EACH		5.41
<NONE>; STRAIGHT CIRC DUCTWORK - 250 mm diameter	m		13.93
<NONE>; STRAIGHT CIRC DUCTWORK - 250 mm diameter clips	EACH		5.41
<NONE>; STRAIGHT CIRC DUCTWORK - 315 mm diameter	m		16.38
<NONE>; STRAIGHT CIRC DUCTWORK - 315mm - 500mm diameter clips	EACH		6.50
<NONE>; STRAIGHT CIRC DUCTWORK - 350 mm diameter	m		19.22
<NONE>; STRAIGHT CIRC DUCTWORK - 400 mm diameter	m		24.82
<NONE>; STRAIGHT CIRC DUCTWORK - 450 mm diameter	m		28.01
<NONE>; STRAIGHT CIRC DUCTWORK - 500 mm diameter	m		30.97
<NONE>; STRAIGHT RECT DUCTWORK -Sum of two sides 200 mm	m		28.83
<NONE>; STRAIGHT RECT DUCTWORK -Sum of two sides 250 mm	m		30.48
<NONE>; STRAIGHT RECT DUCTWORK -Sum of two sides 300 mm	m		32.11
<NONE>; STRAIGHT RECT DUCTWORK -Sum of two sides 350 mm	m		33.21
<NONE>; STRAIGHT RECT DUCTWORK -Sum of two sides 400 mm	m		34.28
<NONE>; STRAIGHT RECT DUCTWORK -Sum of two sides 450 mm	m		35.39
<NONE>; STRAIGHT RECT DUCTWORK -Sum of two sides 500 mm	m		33.86
<NONE>; STRAIGHT RECT DUCTWORK -Sum of two sides 550 mm	m		37.55
<NONE>; STRAIGHT RECT DUCTWORK -Sum of two sides 600 mm	m		38.09
<NONE>; STRAIGHT RECT DUCTWORK -Sum of two sides 650 mm	m		39.20
<NONE>; STRAIGHT RECT DUCTWORK -Sum of two sides 700 mm	m		40.83
<NONE>; STRAIGHT RECT DUCTWORK -Sum of two sides 750 mm	m		42.46
<NONE>; STRAIGHT RECT DUCTWORK -Sum of two sides 800 mm	m		43.00

CONTRACTORS GENERAL COST ITEMS

Generally

Generally; Sundries	Item	1.30

BASIC PRICES : LABOUR, PLANT AND MATERIALS

U:MECHANICAL SERVICES - VENTILATION/AIR CONDITIONING SYSTEMS : WORKS OF ALTERATION/SMALL WORKS/REPAIR
ELECTRIC MOTORS

ELECTRIC MOTOR SPARES & ACCESSORIES

SPARES/REPLACEMENT PARTS; FAN ELEC12506_D90FAN - NO LONGER USED	Item	18.36

ELECTRIC MOTORS - THREE PHASE

3 PHASE 50HZ ALUMINIUM. FOOT & FLANGE MOUNTING. FRAMES 63-250. 0.18KW-55KW; D63 0.18KW FOOT MTG 3PHASE 50HZ ALUM MOTOR D63-.18KW1500FM - NO LONGER USED	Item	96.64

INDUSTR. PUMPS

ASPEN PUMPS

<NONE>; UNIVERSAL PERISTALTIC PUMP	nr	93.46

L/GIANT COND PUMPS

<NONE>; VCC20S AUTO CONDENSATE PUMP COMPACT TANK	nr	113.00
<NONE>; VCL24S AUTO CONDENSATE PUMP LARGE TANK	nr	137.00
<NONE>; VCL45S AUTO CONDENSATE PUMP LARGE TANK	nr	293.00
<NONE>; VCMA20S AUTO CONDENSATE PUMP MEDIUM TANK	nr	128.00